Introduction to Heating, Ventilation, and Air Conditioning
How to Calculate Heating and Cooling Loads

by Thomas E. Mull, P.E.

BNP
Business News Publishing Company
Troy, Michigan

Library of Congress Cataloging in Publication Data
Mull, Thomas E.
Introduction to HVAC: how to calculate heating and
cooling loads / Thomas E. Mull
p. cm.
Includes index.
ISBN 0-912524-84-7
1. Heating load--Mathematics   2. Cooling load--
Mathematics. I. Title.
TH7015.M85   1995   95-18689
697'.00151--dc20   CIP

Administrative Editor: Joanna Turpin
Editors: Barbara A. Checket-Hanks, Carolyn Thompson
Art Director: Mark Leibold

Printed in the United States of America
7 6 5 4 3 2 1

Dedication

This book is dedicated to my wife Norma, my father Elmer, and my late mother Marie Mull. Without their inspiration and support, this book would not have been possible.

About the Author

Thomas E. Mull has more than 20 years of experience as a mechanical engineer. His extensive experience includes energy management, heating and air conditioning, utilities, plumbing, and fire protection. His work has involved a wide range of facilities, including medical, institutional, commercial, industrial, educational, retail, and governmental. During his work, Mr. Mull has traveled internationally and has been involved with projects in Europe and the Pacific Rim. He is the founder and president of Technology Resource Consulting Engineers, Inc., a St. Louis, Missouri-based consulting engineering firm.

Thomas Mull holds bachelor's and master's degrees in mechanical engineering from the University of Missouri-Rolla. He has performed engineering research into passive solar energy and the interaction of building lighting and environmental conditioning systems. Mr. Mull has also taught engineering-related courses at St. Louis Community College.

Mr. Mull is a registered mechanical engineer in several states, including Missouri and Illinois. He also is a certified energy analyst for the state of Missouri and a member of the American Society of Mechanical Engineers, as well as the American Society of Heating, Refrigerating and Air-Conditioning Engineers.

His Internet E-mail address is Mullt.mbr@asme.org.

Preface

This manual is intended as an introduction to heating, ventilating, and air conditioning (hvac) design. The major emphasis is on load calculations for system sizing, although other topics are also included. The methods and procedures presented here are based upon those recommended by the American Society of Heating, Refrigerating and Air-Conditioning Engineers (ASHRAE).

Many books and manuals on hvac design that are available today are intended primarily as handbooks or reference books for degreed engineers practicing hvac design. This manual is intended for engineers not already familiar with hvac design, non-degreed people who are currently working in the hvac design field, people who are considering entering the field, or people in related fields who have an interest in hvac design. It is intended for readers who have at least completed high school and perhaps some community college level courses. While every attempt has been made to keep the mathematics as simple as possible, readers should have at least completed high school algebra and trigonometry. Readers should also have some background and interest in science and technology.

Table of Contents

CHAPTER 1
INTRODUCTION TO HVAC DESIGN _____ 1

CHAPTER 2
BASIC SCIENTIFIC PRINCIPLES _____ 9

CHAPTER 3
CLIMATIC CONDITIONS _____ 19

CHAPTER 4
BUILDING HEAT TRANSMISSION SURFACES _____ 33

CHAPTER 5
INFILTRATION AND VENTILATION _____ 41

CHAPTER 6
HEATING LOADS _____ 47

CHAPTER 7
EXTERNAL HEAT GAINS AND COOLING LOADS _____ 57

CHAPTER 8
INTERNAL HEAT GAINS AND COOLING LOADS _____ 87

CHAPTER 9
HVAC PSYCHROMETRICS _____ 107

CHAPTER 10
OVERVIEW OF HVAC SYSTEMS _____ 129

References _____ 149

Appendix A
Abbreviations, Symbols, and Conversions _____ 151

Appendix B
Thermal Properties of Selected Materials _____ 155

Index _____ 165

Chapter 1

Introduction to Hvac Design

When designing and selecting a heating, ventilating, and air conditioning (hvac) system for a building, whether it be residential, commercial, or industrial, there are a number of general considerations the designer should bear in mind:

- The system should provide a comfortable environment for the occupants, in addition to meeting any special process requirements, such as special humidity control.

- The system should operate efficiently, without wasting energy or being unnecessarily expensive to operate.

- The designer must also be mindful of indoor air quality (IAQ) and attempt to reduce the possibility of indoor air pollutants.

With these basic requirements of the hvac system in mind, the designer must select the type of system that best suits the needs of the building and its occupants. The designer should consider the relative advantages and disadvantages of each type of system.

Providing a comfortable indoor environment

Air conditioning is the simultaneous control of temperature, humidity, air movement, and quality of air within a space. The hvac system attempts to maintain desired indoor conditions despite changes in outdoor weather, indoor activity, or usage of the space.

For example, a system is typically required to maintain an indoor temperature between 70° and 78°F whether the outdoor temperature is 0° or 100°F. The system also must maintain the desired conditions when the space is empty, when it is occupied with many people, or when there are many interior heat gains, such as lights and equipment. It may also be necessary for the system to maintain an indoor relative humidity between 30% and 50%.

The use of the space further determines the requirements of the hvac system; for example, a computer room may require special temperature and humidity conditions, while an office may only require comfortable conditions for the occupants.

For our purposes, comfort may be defined as any environmental condition which, when changed, makes a person uncomfortable. We are primarily concerned with air temperature, humidity, air movement, and IAQ within the space. Individual comfort depends on how fast the body is losing heat. People must be kept comfortable regardless of the clothing they are wearing or what activity they are performing.

The objective of the hvac system is to assist the body in controlling its cooling rate in both summer and winter outdoor conditions and the range of conditions in between. In the summer, it is necessary to increase the body's cooling rate; in winter, it is necessary to decrease the body's cooling rate.

The human body loses or gains heat from its surroundings by three means: sensible heat loss or gain, latent heat loss, and radiant heat transfer.

Sensible heat loss or gain occurs as a result of direct contact of the skin with the surrounding air. The amount of sensible heat transferred depends on the difference in temperature between the skin and the surrounding air. As the difference in temperature between the skin and the air decreases, the amount of sensible heat lost or gained by the body decreases. When the surrounding air temperature is around 70°F, most people lose sensible heat at a rate at which they are comfortable. If the air temperature rises to 80°F, sensible heat loss from the body approaches zero, and people begin to feel hot. If the air temperature rises above 80°F, the human body will begin absorbing or gaining sensible heat.

The human body also loses heat through *latent heat loss* or perspiration. Latent heat is lost by the body as moisture evaporates, or changes, from a liquid to a vapor. As the moisture evaporates, heat is absorbed from the body and is carried away by the vapor. At 80°F, sensible heat loss by the body is negligible; however, latent heat loss from the body can be around 400 Btuh (Btu/hour). As the surrounding air temperature increases, the body is able to continue losing heat through increased perspiration. If the body did not give off latent heat, it would overheat whenever the ambient air temperature rose above 80°F.

Radiant heat transfer does not require direct contact between the body and the object with which the body is exchanging heat. An example of radiant heat gain is the feeling of warmth you experience when the sun is shining on your face on a cold sunny day. There is no direct contact between the sun and the skin, but heat is still received from the sun. In order for radiant heat exchange to occur, the body must "see" the object with which it is exchanging heat; it must not be blocked or shaded by another object.

Energy conservation

The availability, cost, and probable consumption of various fuel types are important considerations in the selection of an hvac system. Hvac systems frequently use more than 60% of the energy consumed in buildings; therefore, it is important for systems to be selected and designed with energy conservation in mind. As can be observed from Figure 1-1, hvac systems typically consume more than 75% of the energy used in office buildings.

Of course, this percentage varies with location or climate. Consumption percentages also may vary slightly with building usage; however, you can easily see that space conditioning consumes the largest amount of energy.

Figure 1-2 indicates relative energy consumption for educational buildings. Again, you can see that

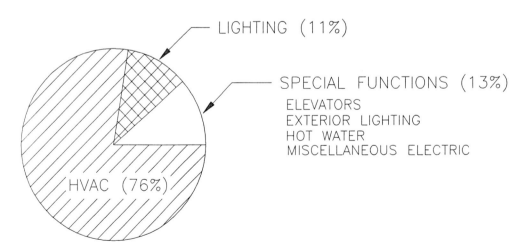

Figure 1-1. Energy consumption in office buildings.

the hvac systems are the major consumers of energy. Although the percentages may vary slightly for different climatic zones, hvac systems account for about 65% of the energy consumed in educational buildings.

Figures 1-3 and 1-4 show that the hvac systems of long-term care facilities, such as nursing homes and hospitals are also large consumers of energy.

While these are a few typical examples, it may be readily seen that hvac systems are the major

consumers of energy in most buildings. Unfortunately, most hvac systems are designed and equipment selected to produce the desired space conditions at minimum first cost, or installed cost. Operating costs are frequently secondary.

Hvac systems are frequently designed using excessively conservative methods, resulting in oversized systems. Oversized systems can be as great a problem as undersized systems. Most systems operate at maximum efficiency at or near their maximum load or capacity. Hvac systems are also frequently sized based on peak heating

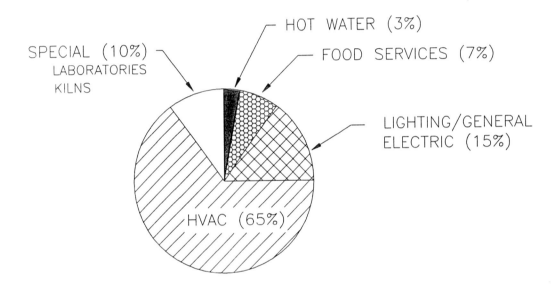

Figure 1-2. Energy consumption in educational buildings.

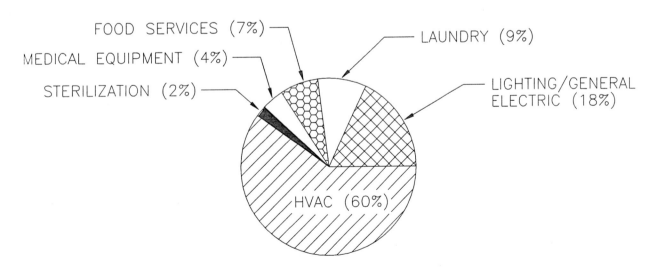

Figure 1-3. Energy consumption in long-term care facility.

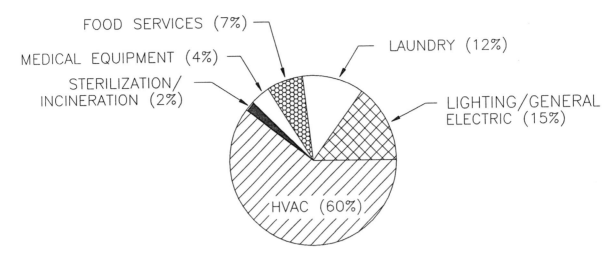

Figure 1-4. Energy consumption in hospitals.

and cooling loads that occur, at most, only a few hours each year. During most of the service life of an hvac system, the systems are operating at considerably less than peak design and efficiency.

The question of energy conservation in hvac systems is more of an economic question rather than a technical question. When considering the economic costs of hvac systems, it is best to look at systems on the basis of life cycle cost, or the total cost to own and operate for the entire life of the system. First cost is only one factor in the economic equation.

In addition to economic considerations, there are code considerations affecting hvac systems. Many codes and standards set maximum energy consumption budgets for buildings in Btu per square foot of floor space per year. It is possible to deviate from these energy budget guidelines in the design of a building if it can be shown that the deviation will result in no additional energy usage. Any combination of building design, construction materials, and hvac system types may be used so long as the maximum budget figure is not exceeded. The most widely used energy conservation standards are published by the American Society of Heating, Refrigerating and Air-Conditioning Engineers (ASHRAE). ASHRAE standards provide minimum guidelines for energy conservation design and operation. There are two types of energy standards:

- *Prescriptive standards* specify the materials and methods for the design and construction of buildings.

- *Component performance standards* set requirements for each component, system, or subsystem within a building.

When designing hvac systems for buildings, consult ASHRAE Standard 90, *Energy Conservation on New Building Design* or ASHRAE Standard 100-P, *Energy Conservation in Existing Buildings*.

Indoor air quality

It is estimated that people in the United States spend 65% to 90% of their time indoors. With so much time spent indoors, the quality of the indoor air we breathe has a significant effect on our health. Unfortunately, the indoor air we breathe is frequently much more polluted than outdoor air. Generally speaking, if more than 20% of a building's occupants complain of symptoms, such as headaches, breathing problems, excessive colds, etc., and no specific cause can be found, the building is probably suffering from *sick building syndrome*.

The National Institute of Occupational Safety and Health (NIOSH) has investigated several hundred buildings and determined that 50% of occupant complaints are due to building material contaminants and 11% are due to outdoor contaminants. It is estimated that the number of "sick" buildings in the United States is between 20% and 30% of all existing buildings. Health complaints include itchy eyes, headaches, rashes, nausea, chronic fatigue, and respiratory problems. When a building has sick building syndrome, occupant productivity can drop off as much as 5%.

Construction materials such as paints, plastics, carpeting, wall coverings, etc., are major sources of indoor air pollution. Approximately 20% of all commercial type buildings contain asbestos, and formaldehyde is used in over 3,000 building products. Accumulated dust and moisture can become environmental niches that promote the growth of microorganisms or molds. These can be drawn into the air stream and circulated throughout the building by the air distribution system. Environmental niches combine organisms with near optimum temperature, humidity, and moisture to cause concentrations to grow very rapidly. An example is Legionnaires' disease, which is associated with standing water. Another common and obvious source of indoor air pollution is cigarette smoke. Since most air distribution systems recirculate air within the building, isolation of smokers and other sources of pollution is not very effective.

Although IAQ is an important consideration when selecting and designing hvac systems, designers often neglect to give it proper attention. System designers are frequently so concerned about temperature control and installed cost they tend to forget about IAQ. In some variable air volume systems, the reduced need for cooling can reduce the amount of airflow to a space to such an extent that a stuffy and possibly unhealthy environment can result. Air filtration is not always effective in reducing indoor air pollution, particularly gaseous air pollution.

When designing an hvac system, the designer should always be aware of any pollution resulting from materials, systems, equipment, or processes within a building. The design should attempt to minimize indoor air pollution. Many building mechanical codes, such as the BOCA Code, prescribe minimum outdoor air requirements for buildings in order to control indoor air pollution. ASHRAE Standard 62 provides guidelines for minimum outdoor ventilation brought into a building in order to dilute indoor air pollutants. In addition to bringing outdoor air into a building, which may also contain some pollutants, steps can be taken in the design of the hvac system to minimize indoor pollutants. Electronic filters and other types of high efficiency filters can remove more than 90% of airborne particles; however, they cannot remove gaseous contaminants. In order to reduce indoor air pollution, the designer should consider the following points:

- Provide differential pressure control within the building to reduce contamination from one area to other spaces (e.g., a smoking room).

- Design the air distribution system to exclude materials that promote contaminants, such as non-metallic duct linings.

- Avoid standing water in condensate drains, equipment, ducts, etc.

- Install efficient upstream filtration.

- Provide adequate space for maintenance of all equipment.

- Fully balance air systems to provide design airflow rates.

- Avoid locating ceiling supply and return grilles in close proximity to each other, as this may cause air to "short circuit."

- Locate outdoor air intakes such that there will be no re-entry of exhaust fumes from toilet exhausts, plumbing, vent stacks, cooling towers, vehicle exhaust, etc.

- Design air systems so they are easily balanced.

- Pay particular attention to reduced airflow rates in variable air volume systems.

Basic hvac system types

There are several major heating, ventilating, and air conditioning system types in widespread use today: air systems, hydronic and steam systems, and unitary type systems. Most current systems fall into one of these categories or some combination thereof. Each system has advantages and disadvantages.

Air systems

Air type hvac systems provide heating, ventilation, and cooling by circulating air through spaces or rooms in order to maintain the desired conditions in the space. A typical air system generally consists of a central air-handling unit with heating and cooling capabilities and an air distribution system. The air distribution system generally consists of supply air ductwork to the space or room, return air ductwork from the space back to the air-handling unit, and air distribution devices in the room, such as grilles and registers, to provide desirable air patterns within the room. Heating of the air in the air-handling unit (AHU) is usually accomplished with hot water coils, steam coils, gas furnaces, or electric heating coils. Cooling and dehumidification are generally accomplished by passing the air through direct expansion refrigerant coils, chilled water cooling coils, or by the introduction of outdoor air when conditions are favorable.

Air systems offer a number of advantages over other types of systems, including the following:

- Central location of major equipment components

- Ventilation provided directly to the space

- Easy to use outdoor air for cooling when conditions are favorable (economizer)

- Many variations to choose from, i.e., multi-zone, variable air volume, dual duct, etc.

Air systems also have a number of disadvantages when compared to other systems, including the following:

- Central air-handling unit (AHU) and ductwork can require large amounts of space.

- It is difficult to efficiently provide heating to one space while providing cooling to another space served by the same system.

- Airflows into spaces need to be properly balanced to obtain the design airflow rates.

- Air systems may have the tendency to overcool and cause draftiness in the space.

- These systems may be quite noisy.

Hydronic and steam systems

Hydronic systems circulate hot water for heating and chilled water for cooling. Steam heating systems are similar to heating water systems except that steam is used as the heating medium instead of water.

Hydronic heating systems generally consist of a boiler, hot water circulating pumps, distribution piping, and a fancoil unit or radiator located in the room or space. Steam heating systems are similar except that no circulating pumps are required.

Hydronic cooling systems consist of a water chiller, circulating pumps, distribution piping, and a fancoil unit in the space or room. Chilled water systems also usually require a cooling tower or some other device to reject heat to the outdoors.

Hydronic systems have several advantages over other types of systems, including the following:

- Individual room temperature control is easily achieved.

- Same distribution piping system can sometimes be used for heating water and chilled water if simultaneous heating and cooling is not required.

- Distribution piping requires considerably less space than ductwork.

- System components have a long service life, 15 to 20 years.

- Heating is usually supplied at the perimeter of the building where it is most needed.

Hydronic systems also have several disadvantages, including the following:

- Difficult to provide adequate ventilation to rooms

- Difficult to take advantage of outdoor air cooling when conditions are favorable (economizer)

- Fancoil units in space can be noisy

- Elaborate condensate drainage system may be required for individual fancoil cooling units

- Usually not economical for smaller buildings
- Fancoil units take up space in the room
- May be difficult to control humidity levels in the rooms
- May freeze when used in air-handling units with outdoor air

Unitary systems

Unitary systems are self-contained units that heat and/or cool a single space. Each space may have its own unitary system. The units produce heating and/or cooling with the application of the proper energy source, usually electricity.

Unitary cooling units are usually completely self-contained and include cooling coils, compressors, refrigerant piping, condenser coils, and controls. These cooling units need to be located near an outside wall or roof so the refrigeration system can reject heat to the outdoors. Unitary heating is usually accomplished with electric heating coils or by a gas furnace within the unit. Heating is occasionally accomplished with a heat pump cycle.

Examples of unitary equipment include window air conditioners, through-the-wall air conditioners and heat pumps, package rooftop air conditioners, electric baseboard radiators, and water-source heat pumps.

Unitary systems have several advantages over other types of systems, including the following:

- Lower installed costs
- Individual control in each room
- Units installed on perimeter of building can easily bring ventilation air into building
- Easy to use temperature controls
- Easy to service and maintain

Unitary systems have several disadvantages, including the following:

- Units with refrigeration compressors are usually quite noisy.
- Precise temperature and humidity control is

difficult if not impossible.

- Cooling units usually must be located near an outside wall or on the roof in order to reject heat to the atmosphere.
- Units can be unsightly and aesthetically unpleasing.
- Condensate drainage can be a problem with some cooling units.
- Many units can be very time consuming to maintain.
- Compressors have a relatively short service life, 5 to 10 years.

System selection criteria

The decision as to which type of system is best suited for a particular application is based upon several selection criteria. No one type of system will be best suited for every application, and each application must be considered on an individual basis. Some of the major considerations are as follows.

Installed cost: Some building owners are primarily concerned with initial installed costs. Some may have specialized operating requirements, such as a computer room, while others are more concerned about operating costs and are willing to pay for a higher performance system.

Operating cost: Some systems are inherently more costly to operate than others.

Individual temperature control: Some systems can more readily provide individual room or zone temperature control than others.

Adequacy of ventilation: Some systems by their very nature provide better ventilation than others.

Specialized temperature and humidity control: Some areas, such as computer rooms and special manufacturing processes, frequently require precise temperature and humidity control.

Maintenance requirements: Consideration should be given to the capabilities of the maintenance staff.

Aesthetics: Some owners are very conscious of the appearance of the building, both inside and outside.

Space requirements: It is necessary to consider space for maintenance, and how much space is likely to be available for the system.

Availability of equipment: Some equipment is "off the shelf," whereas others may have long lead times to manufacture or may even have to be custom built.

Climate: Some systems, such as air source heat pumps, are better suited for one climate than another.

Availability of energy sources: Some fuel sources, such as natural gas, are not always available at the building site.

Acoustics: It is always important to consider the amount of noise and vibration generated by various hvac system types. Some applications are more sensitive than others.

Complexity: The system selected should be able to adequately handle the requirements of the building, yet not be so overly complex that it is difficult to have the desired performance.

Chapter 2

Basic Scientific Principles

Before a heating and air conditioning system can actually be designed, the heating and cooling loads, or requirements, must be determined. The design of hvac systems is actually the application of science and scientific principles, along with technology, to achieve a desired result. In this case, the desired result is control of the indoor environment.

To understand how hvac systems are designed and how they actually operate, it is first necessary to understand the basic underlying scientific principles. Among the physical sciences that are essential to hvac design are the fundamental sciences of thermodynamics, heat transfer, and fluid mechanics.

Heat and work

Heat is a form of energy that, when added to a substance, causes that substance to rise in temperature, fuse, evaporate, or expand. Heat itself cannot be seen or observed; it is only possible to observe the effects of heat. The primary effect of adding heat to a substance is an increase in the vibration of the molecules of the substance, which causes the temperature of that substance to rise. Heat always flows from a hotter area to a colder area, never in reverse.

The standard unit of measurement for heat is the British thermal unit (Btu). A Btu is the amount of heat required to raise the temperature of 1 lb of water 1°F. Another commonly used measure-

ment of heat in SI (metric) units is the kilowatt-hour (kWh), which is equal to 3,412 Btu. (Conversion factors for units of measurement commonly used in the hvac industry can be found in Appendix A.)

Heat occurs in two forms. **Sensible heat** changes the temperature of a substance. An example of sensible heat is a substance such as water, air, or a solid increasing or decreasing in temperature. **Latent heat** is heat that produces a change in state, or phase change, of a material. Examples of latent heating include evaporation, condensation, melting, and boiling. In each case, heat is added to or removed from the substance, but the temperature of the material remains constant as the phase change occurs.

Another important concept is **specific heat**. The specific heat is the amount of heat required to raise the temperature of 1 lb of a substance 1°F. The units for specific heat are Btu/lb. Specific heat is a measure of a substance's ability to absorb and store heat. Some substances require more heat to raise their temperature than others. For example, the specific heat of water is 1.0 Btu/lb, while the specific heat of air is 0.24 Btu/lb. It takes almost four times as much heat to raise one pound of water 1°F than it does to raise one pound of air 1°F.

Work is another important concept in hvac design. Work is a transient form of mechanical energy. Work is done on an object whenever a force is exerted on the object, and the object is displaced or accelerated in the direction of the force. The commonly used units for work are foot-pounds per minute (ft-lb/min).

Power is the time rate at which work is done. It is measured in horsepower (hp).

Examples of work commonly found in hvac systems include fans circulating air, pumps circulating liquids, and compressors compressing gases.

Thermodynamics

Central to the study of heating and air conditioning is the concept of a **control volume**. A control volume can be any part of the system in which the flow of a energy into and out of an area is studied. An example might be a room that is part of the system as a whole, but it is necessary to know the net heat flowing into and out of that room. The room is the control volume; the walls, ceiling, and floor are boundaries of the control volume. It may be necessary to examine the heat flowing out through the walls in order to determine the amount of heat that must be added to maintain the room temperature.

Another control volume may be the air-handling system itself. Heat is added to the control volume by a heating device such as a coil. Work or mechanical energy is added by a fan; energy flows out of the control volume through air grilles and registers.

Another control volume is the heating coil. If it is a steam heating coil, heat flows into the control volume (room) in the form of steam and out as heat is transferred to the airstream.

A control volume is used whenever it is necessary to isolate a particular component, in order to analyze the flow of energy into and out of the control volume or component.

Energy can have several different forms. Heat and work have already been discussed. Other forms of energy include potential energy (from the release of a spring or pressure); kinetic energy (objects with mass in motion; the energy that moves the mass is from an applied outside force); and chemical energy (from the combustion of natural gas and gasoline). Potential energy may also result from the elevation of a mass.

Heat is the lowest form of energy. All other forms of energy degrade to heat. Mechanical energy (work) and kinetic energy degrade to heat through function. Pressure degrades to heat through friction. Chemical energy is transformed to heat via a chemical reaction, such as burning.

The **First Law of Thermodynamics** (the conservation of energy law) states that the energy flowing into a control volume is equal to the energy flowing out of a control volume less the energy stored within the control volume. A simple mathematical expression for the First Law of Thermodynamics is:

$$\text{Energy in} = \text{Energy out} - \text{Energy stored}$$

Equation 2.1

This is shown schematically in Figure 2-1, where the dashed box indicates the boundaries of the control volume. Energy can cross the system boundaries in any form (e.g., heat, work, chemical energy, or potential energy).

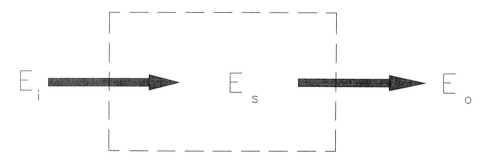

$$E_i \qquad E_s \qquad E_o$$

Figure 2-1. Control volume for First Law of Thermodynamics.

Consider a typical hvac application. Suppose a room (a control volume) has a wall exposed to cold outside temperatures and a radiator to supply heat. Energy in the form of heat flows out of the room, or control volume, and heat energy flows into the room from the radiator. If the heat flowing out of the room through the wall (heat loss) is equal to the heat flowing into the room (heat gain), no energy is stored, and the room temperature remains constant.

If, however, the heat supplied by the radiator is greater than the heat loss, energy is stored in the room in the form of heat. This addition of stored heat results in an increase in room air temperature. Conversely, if the heat loss is greater than the heat gain from the radiator, a net heat loss occurs, resulting in negative heat storage and a decrease in the room air temperature.

The First Law of Thermodynamics and the concept of a control volume are applied over and over in hvac design. A control volume can be an entire building, an entire system (such as an air conditioning system), or an individual component in the system (an evaporator coil). In each case, the First Law of Thermodynamics is used to perform an energy balance on the control volume in question. The usefulness of the First Law becomes obvious in later chapters, when the calculation processes for determining heating and cooling loads are studied.

Forms of heat

Sensible heat is most familiar to us. A change in sensible heat content causes a change in a substance's temperature and is measured by an ordinary thermometer. The temperatures measured are called **dry bulb temperatures**.

Latent heat is associated with a phase change of a substance. A phase change may be defined as a quantity of a substance that is homogeneous throughout. For example, water may be in the liquid phase, vapor phase (steam), or solid phase (ice).

If water is in the vapor phase, it is completely steam with no liquid or solid water present. A change of phase can be from a liquid to a vapor or gas (evaporation); from a vapor or gas to a liquid (condensation); from a liquid to a solid (solidification); or from a solid to a liquid (melting). A substance may also change directly from a solid to a gas or vapor (sublimation). In each case, heat is either added to or removed from the substance as it changes from one phase to another.

During the phase change, the temperature of the substance remains constant as heat is added or removed, until all of the substance completely changes phase. For example, the temperature of boiling water remains constant at 212°F as latent heat is added, until all of the water boils away and becomes steam.

Enthalpy

Enthalpy is a measure of the total energy content of a substance. It is a thermal property of a substance, consisting of the internal energy of the substance plus energy in the form of work due to flow into or out of a control volume. The energy due to flow is the product of the pressure times the volume.

Internal energy, which is the major component of enthalpy, is energy associated with the motion and configuration of the interior particles of a substance, primarily due to temperature resulting from the addition of heat.

Enthalpy is frequently used in hvac calculations, although it is actually the total heat content of a substance. The flow work component in hvac applications is usually ignored, because it is relatively small. Hvac calculations are primarily concerned with the total heat of air, including any suspended water vapor in the form of latent heat. When the term enthalpy of the air is used, it is really referring to the total heat content of the air-vapor mixture.

Specific heat

As stated previously, specific heat is the amount of heat required to raise the temperature of one pound of a substance 1°F. Specific heat may be expressed as:

$$q = (mc)(T_2 - T_1)$$

Equation 2.2

where:

- q = quantity of heat added or removed from the substance (Btu)
- m = mass of the substance (lb)
- c = specific heat of the substance (Btu/lb-°F)
- $T_2 - T_1$ = temperature difference (°F)

Since water is a common substance, it has arbitrarily been chosen as the standard measure of specific heat. In the English unit system, water has a specific heat of 1.0. The heat required to raise one pound of water 1°F is one Btu. Therefore, the specific heat of water is 1.0 Btu/lb-°F.

In general, the specific heat of a substance remains constant. Specific heat does vary slightly with temperature, but for hvac calculations it may be assumed a constant.

Heat transfer

Heat transfer involves the transfer or flow of energy in the form of heat. Whenever a temperature difference (or gradient) exists from one side of a substance to another or when a temperature difference exists between two bodies, heat transfers from the high-temperature region to the low-temperature region.

The study of heat transfer attempts to predict the rate at which heat is transferred through a substance or from one body to another. Unlike thermodynamics, heat transfer does not always involve the transfer or flow of mass in order for heat to flow from the high-temperature region to the low-temperature region.

There are three methods by which heat is transferred through a substance or between bodies: conduction, convection, and radiation.

Conduction

Conduction heat transfer occurs through, and by means of, intervening matter; it does not involve any obvious motion of the matter. It is a mechanism of internal energy exchange from one part of a substance to another part by the exchange of energy between molecules of the substance. More simply expressed, the molecules physically touch each other and transfer heat as they touch.

Suppose a thick homogeneous concrete wall has a hot plate applied to one side and a cold plate applied to the other side. Assume also that the hot and cold plates have been applied for a long period of time. Heat would flow through the concrete wall from the hot plate to the cold plate at a constant rate. The rate of heat flow would be a function of the wall material and its thickness.

Concrete offers some resistance to heat. Other materials transmit or conduct heat more readily. For example, glass is frequently used for cookware, because it conducts heat rather quickly and easily. Metals are also excellent heat conductors.

On the other hand, wood and polystyrene ("Styrofoam") are poor conductors of heat and are frequently used as insulators. Thermal conductance is a measure of the ease with which heat travels through material of a specified thickness. Knowing the thermal conductance of a given material determines the rate at which heat is transferred. The following equation may be used to predict the rate at which heat is conducted through a given material of a given thickness:

$$q = (CA)(T_2 - T_1)$$

Equation 2.3

where:

- q = rate of heat transfer (Btuh)
- C = thermal conductance of material of specified thickness (Btuh/sq ft-°F)
- A = surface area of the material (sq ft)
- $T_2 - T_1$ = difference in temperature (°F)

This equation is used to predict the steady-state conduction of heat. In other words, it is assumed that the temperature difference $(T_2 - T_1)$ has existed for a long period of time and the rate of heat transfer is a constant (not changing with time).

Example 2-1.
Assume that 100 sq ft of 1/2-in. thick gypsum board (C = 2.2 Btuh/sq ft-°F) has a temperature of 70°F on one side and 10°F on the other side. At what rate is heat conducted through the gypsum board?

Use Equation 2.3 to determine the rate:

q = (2.2 Btuh/sq ft-°F) (100 sq ft) (70°F - 10°F) = 13,200 Btuh

Thermal conductance is a measure of the ease with which a material of a given thickness transmits heat. The reciprocal of the conductance gives the resistance to heat flow (R). The higher the R value, the greater the resistance to the flow of heat. The equation for resistance is as follows:

$$R = \frac{1}{C}$$

Equation 2.4

Thermal conductance is given for a material of a given thickness. Tables often give the conductance per unit thickness, which is known as thermal conductivity of the material. The equation for thermal conductance is as follows:

$$C = \frac{k}{x}$$

Equation 2.5

where:
k = thermal conductivity (Btuh/sq ft-°F-in. or Btuh/sq ft-°F)
x = material thickness (in. or ft)

Note: Care must be taken to ensure that all units are consistent.

Example 2-2.
What is the thermal conductance of 1/2-in. gypsum board that has a thermal conductivity (k) of 0.0926 Btuh/sq ft-°F-in.?

Use Equation 2.5 to determine the conductance. First, convert the material thickness (x) to feet:

$$x = \left(1/2 \text{ in.}\right)\left(\frac{1 \text{ ft}}{12 \text{ in.}}\right) = 0.0417 \text{ ft}$$

$$C = \frac{k}{x} = \frac{0.0926}{0.0417} = 2.22 \text{ Btuh/sq ft-°F}$$

Thermal conductance and other thermal properties for common building materials are given in Appendix B.

Convection

Convection is a heat transfer mechanism that occurs by means of a current or circulation of a fluid (liquid or gas). Heat is transferred within the fluid by mixing the molecules within the fluid.

An example of convection heat transfer is cold air flowing over a warm surface. The air flowing closest to the surface is heated as it comes in contact with the surface, and the air heated at the surface mixes with the free airstream.

Convection heat transfer can be due to either forced or natural flow. In the example just given, the flow of air over the surface could be forced by a fan. An example of natural convection is a cold fluid next to a warm vertical surface. As the fluid next to the surface is heated, the natural buoyancy forces cause the warmed fluid to rise due to the change in density of the fluid. This causes a natural flow of fluid past the surface. Conversely, if the fluid is cooled by the surface, the fluid next to the surface will drop.

In both forced and natural flow, the fluid molecules right at the surface are not moving; the fluid slightly farther away *is* moving. Layers of fluid molecules still farther away are moving even faster until the mean free stream velocity of the fluid is reached. Fluid mechanics will be discussed later in this chapter.

The equation for convection heat transfer is known as **Newton's Law of Cooling** and may be expressed as follows:

$$q = (hA)(T_s - T_f)$$

Equation 2.6

where:

q = rate of heat transfer (Btuh)
h = convection heat transfer coefficient (Btuh/sq ft-°F)
A = surface area (sq ft)
T_s = surface temperature (°F)
T_f = free stream fluid temperature (°F)

Values for the convection heat transfer coefficient (h) are determined experimentally. In a manner similar to the resistance for thermal conductance, the resistance for the convection coefficient is as follows:

$$R = \frac{1}{h}$$

Equation 2.7

(See Table 4-1 in Chapter 4 for air film resistances.)

Radiation

Radiated heat is transferred via electromagnetic waves. In order for radiation heat transfer to occur between two bodies, the two bodies do not need to be in direct physical contact. It is also not necessary that there be a medium through which heat is transferred, as radiation can pass through a complete vacuum. An example of radiation is the sun heating the surface of the earth.

Objects exchanging radiation must "see" each other and not be shaded. Radiation travels in a straight line at the speed of light. The intensity of radiation falling on a surface is a function of the angle at which the radiation strikes the surface. The angle is called the **angle of incidence**. The more direct the radiation, the more intense it is.

When radiant energy strikes a surface, part of the energy is absorbed, part is reflected away, and part is transmitted through the surface, as shown in Figure 2-2.

Radiation heat transfer is proportional to the difference between the fourth power of the

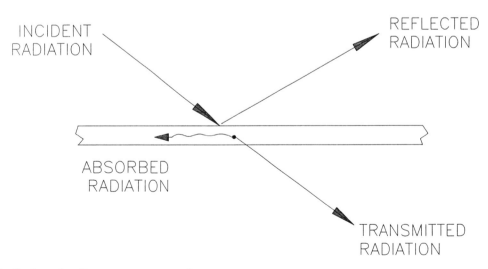

Figure 2-2. Distribution of radiant energy on a surface.

absolute temperatures of the two bodies; it may be expressed as follows:

$$q = (\sigma \, A F_e \, F_s)(T_1^4 - T_2^4)$$

Equation 2.8

where:

q = rate of heat transfer (Btuh)
σ = a constant, 0.1714×10^{-8} Btuh/sq ft-$°R^4$
F_e = absorption factor
F_s = view factor
T_1 and T_2 = absolute temperatures of the bodies in degrees Rankine (R)
A = surface area (sq ft)

Transient heat transfer

Previously, all of the heat transfer examples were steady state, one directional. In other words, the rate of heat transfer was assumed to be constant over a long period of time and in only one direction. In steady-state heat transfer, heat entering an object (or control volume) is equal to the heat leaving, and it is assumed that this has been the case for a long time. In actual hvac applications, this is frequently not the case.

Consider a situation where a concrete wall is shaded from the sun's rays by some external object. Assume that the object is suddenly moved and the sun's rays shine on the outside surface of the wall. The exterior surface immediately begins to increase in temperature; the interior surface does not. There will be a considerable lag between the time the sun's rays initially strike the exterior surface and when the inside surface begins to increase in temperature. The rate of heat transfer through the wall, then, is not constant but varies with time.

The time lag between the application of heat to one surface and when the heat begins to appear on the other side, is due to the material's thermal or heat storage. The difference between the instantaneous heat gain on a building and the actual cooling load the air conditioning system must absorb is illustrated in Figure 2-3.

Note in Figure 2-3 that the instantaneous heat gain peaks sooner and is greater than the actual cooling load. Only where the instantaneous heat and cooling load curves cross is the heat gain equal to the cooling load. Prior to the time when the curves cross, heat gain is greater than the cooling load and heat is stored by the building mass. After the curves cross, the building is no

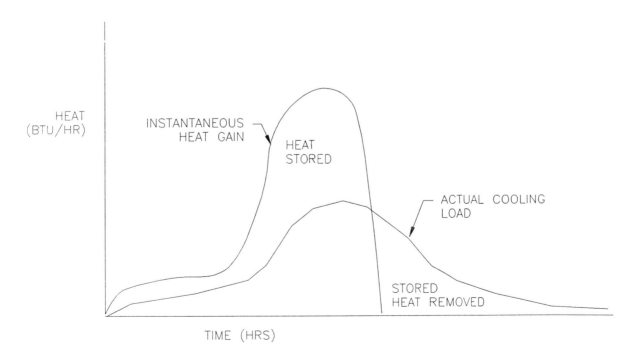

Figure 2-3. The effect of thermal storage on cooling loads.

longer storing heat. The building mass is rejecting heat to the cooling system.

The thermal mass of a building directly affects the time lag between the peak heat gain and peak cooling load, and it inversely affects the intensity of the peak cooling load. The greater the thermal mass of the building, the greater the time lag and the lower the peak intensity of the cooling load. This is illustrated in Figure 2-4.

In some cases, the time lag between the heat gain and the cooling load is so great that the peak cooling load for the air conditioning system occurs at night when the building is empty. This will be discussed again in Chapters 4, 7, and 8.

Fluid mechanics

A fluid may be defined as a substance that deforms continuously when subjected to a shear stress, no matter how small that stress may be. A fluid is a substance that is capable of flowing, having particles that easily move and change their relative position without separation of the mass. Fluids also yield easily to pressure. A fluid may be a liquid, such as water, or it may be a gas, such as air.

Pressure

Any fluid at rest or in motion exerts a perpendicular, or normal, force on a boundary, which is known as pressure. It is also a force exerted on an object submerged in the fluid. Pressure measured relative to a perfect vacuum is called absolute pressure. Gauge pressure includes the absolute pressure plus the surrounding atmospheric pressure, which is normally considered to be 14.7 lb/sq in.:

$$Gauge\ pressure = Absolute\ pressure + Atmospheric\ pressure$$

Equation 2.9

Pressure is usually measured in pounds per square inch (psi), although it may be measured in pounds per square foot, feet of water, and inches of water. One foot of water pressure is the pressure necessary to raise a column of water one foot vertically. For relatively low pressures,

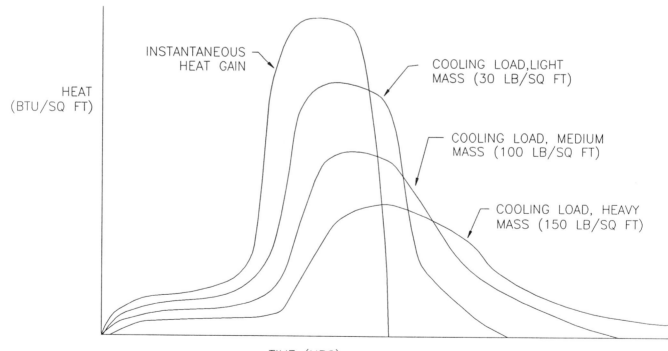

HEAT (BTU/SQ FT)

INSTANTANEOUS HEAT GAIN

COOLING LOAD, LIGHT MASS (30 LB/SQ FT)

COOLING LOAD, MEDIUM MASS (100 LB/SQ FT)

COOLING LOAD, HEAVY MASS (150 LB/SQ FT)

TIME (HRS)

Figure 2-4. The effect of building mass on cooling loads.

measurements are usually given in inches of water column (in. wc) or inches of mercury (in. Hg).

Viscosity

Viscosity is a measure of the ease with which a fluid flows. Viscosity enables a fluid to develop and maintain an amount of shearing stress dependent upon the velocity of the flow and then to offer continued resistance to flow. Energy is lost due to friction in the flowing fluid due to the presence of its viscosity.

Viscosity is caused by the molecular structure of the fluid. In the liquid state, the molecules are packed closely together and the viscosity is due to the cohesiveness of the molecules. As the temperature of a liquid increases, the cohesion decreases and the viscosity decreases.

For a gas, the molecules are spaced much farther apart and the viscosity is due to the activity of the molecules striking against each other. As the temperature increases, the activity of the molecules increases, the collisions increase, and the viscosity of the gas increases.

The continuity equation

According to the continuity equation, which is known as the **Law of Conservation of Mass for Fluid Flow**, no fluid is created or destroyed. The continuity equation is important to the study of the flow of fluids in closed conduits, such as pipes or ducts. The continuity equation states that, for steady flow, the volume flow rate is the same at any location. Mathematically, it is stated as follows:

$$Q = A_1V_1 = A_2V_2 = \text{Constant}$$

Equation 2.10

where:

Q = rate of fluid flow (cfm, gpm, etc.)
A = cross sectional area of conduit (feet)
V = velocity of the fluid (fpm)

This equation assumes steady-state flow, that is, a constant flow rate and no compression in the case of a gas.

The continuity equation is used to relate fluid velocity and flow area at two sections of a closed conduit in which there is steady flow. It is valid for all fluids. It may be observed from the equation that, if the area of the conduit changes from one section to another, the velocity must also change for any given constant flow rate. This assumes no flow into or out of the conduit between the two sections.

Fluid flow friction losses

As we discussed in the section on convection heat transfer, whenever a fluid flows in a closed conduit or along a solid surface, the fluid right next to the surface is actually not moving and has zero velocity. This is due to friction between the flowing fluid molecules and the solid surface. Figure 2-5 shows a typical velocity profile for a fluid flowing in a conduit such as a pipe.

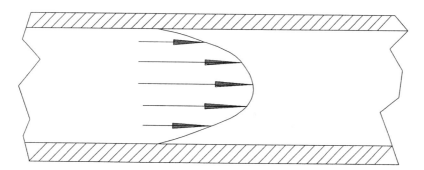

Figure 2-5. Fluid velocity profile.

In Figure 2-5, the arrows represent relative velocities of the fluid, the longer arrows indicating higher velocities and the shorter arrows indicating lower velocities. You can see that while the fluid near the walls of the conduit has a velocity of zero, near the center of the conduit the velocity is much greater. The velocity near the center is known as the **free stream velocity**.

The velocity gradient that exists between the fluid at the walls and the center is due to friction between the walls and the molecules of the fluid and between the molecules themselves. The energy loss due to friction may be expressed as:

$$h_l = \frac{fLv^2}{D2g}$$

Equation 2.11

where:
  h_l = energy (pressure) loss due to friction (feet of water)
  L = length of flow stream (feet)
  D = pipe diameter (feet)
  v = average flow velocity (ft/sec)
  f = a friction factor (dimensionless)
  g = acceleration due to gravity (32.2 ft/sec^2)

The friction factor, f, is a function of the fluid velocity, fluid viscosity, and the roughness of the conduit walls.

Equation 2.11 is known as **Darcy's Equation** and is important in hvac design when considering the flow of air in ducts or fluid flow in pipes.

Chapter 3

Climatic Conditions

When selecting a suitable heating and air conditioning system for a building, it is first necessary to determine the indoor heating and cooling loads, part of which will be attributable to the outdoor climatic conditions for the building location.

This chapter presents detailed climatic data and considerations used when calculating design heating and cooling loads.

Design temperatures, along with other weather data for various locations, are presented in Table 3-1. The weather data was measured at first-order weather stations, airports, and Air Force bases, by trained observers. The recorded values were statistically analyzed and tabulated for a 15-year period.

In Table 3-1, winter and summer design temperatures are given for various percentages, representing frequency risk levels for the dry bulb temperature listed. It is assumed that the frequency level for any given dry bulb temperature will repeat itself in the future. The percentages can be used to ascertain the level of risk that the design temperature will be exceeded.

The winter percentages are for the months of December, January, and February in the northern hemisphere. The three winter months equal a total of 2,160 hours. We may assume that the design dry bulb temperature given in the 99% column will not be exceeded (the temperature will not drop below) 99% of the time or, conversely, will be exceeded only 1% of the time during the three winter months. This would be equal to a total of 22 hours.

For the 97.5% frequency level, the design temperature will be exceeded only 2.5% of the time, or 54 hours. Similarly, the summer percentages and dry bulb temperatures are for the four summer months of June, July, August, and September (2,928 hours) in the northern hemisphere. The wet bulb temperatures are mean coincident temperatures at the given dry bulb temperature.

Mean daily range temperatures are the difference between the average daily maximum and the average daily minimum dry bulb temperatures for the warmest month. These are also tabulated.

When selecting a percentage frequency and the corresponding temperature, it is necessary to consider the effect on the hvac system and the usage of the building. The lower the frequency risk level, the greater the difference will be in the indoor, and outdoor, design temperatures. This will result in an hvac system of greater capacity and greater cost.

The intended use of the building is usually the determining factor in the selection of the frequency risk level. In a building in which it is critical to maintain indoor design conditions, such as a hospital, nursing home, or computer room, the lower frequency risk level may be chosen (i.e., 99% for winter and 1% for summer). A building that is not so critical, such as a retail or office building, may only require a higher frequency risk level (i.e., 97.25% for winter and 2.5% for summer).

Based on the above considerations, the design outdoor temperatures and other weather related data for a given location may be chosen from Table 3-1.

Col. 1	Col. 2	Col. 3	Col. 4	Winter,[b] °F Col. 5		Summer,[c] °F Col. 6			Col. 7	Col. 8			Prevailing Wind Col. 9			Temp. °F Col. 10	
State and Station[a]	Lat.	Long.	Elev.	Design Dry-Bulb		Design Dry-Bulb and Mean Coincident Wet-Bulb			Mean Daily	Design Wet-Bulb			Winter		Summer	Median of Annual Extr.	
	° ′	° ′	Feet	99%	97.5%	1%	2.5%	5%	Range	1%	2.5%	5%	Knots[d]			Max.	Min.
ALABAMA																	
Alexander City	32 57	85 57	660	18	22	96/77	93/76	91/76	21	79	78	78					
Anniston AP	33 35	85 51	599	18	22	97/77	94/76	92/76	21	79	78	78	SW 5		SW	98.4	12.4
Auburn	32 36	85 30	652	18	22	96/77	93/76	91/76	21	79	78	78				99.8	14.6
Birmingham AP	33 34	86 45	620	17	21	96/74	94/75	92/74	21	78	77	76	NNW 8		WNW	98.5	12.9
Decatur	34 37	86 59	580	11	16	95/75	93/74	91/74	22	78	77	76					
Dothan AP	31 19	85 27	374	23	27	94/76	92/76	91/76	20	80	79	78					
Florence AP	34 48	87 40	581	17	21	97/74	94/74	92/74	22	78	77	76	NW 7		NW		
Gadsden	34 01	86 00	554	16	20	96/75	94/75	92/74	22	78	77	76	NNW 8		WNW		
Huntsville AP	34 42	86 35	606	11	16	95/75	93/74	91/74	23	78	77	76	N 9		SW		
Mobile AP	30 41	88 15	211	25	29	95/77	93/77	91/76	18	80	79	78	N 10		N		
Mobile Co	30 40	88 15	211	25	29	95/77	93/77	91/76	16	80	79	78				97.9	22.3
Montgomery AP	32 23	86 22	169	22	25	96/76	95/76	93/76	21	79	79	78	NW 7		W	98.9	18.2
Selma-Craig AFB	32 20	87 59	166	22	26	97/78	95/77	93/77	21	81	80	79	N 9		SW	100.1	17.6
Talladega	33 27	86 06	565	18	22	97/77	94/76	92/76	21	79	78	78				99.6	11.2
Tuscaloosa AP	33 13	87 37	169	20	23	98/75	96/76	94/76	22	79	78	77	N 5		WNW		
ALASKA																	
Anchorage AP	61 10	150 01	114	−23	−18	71/59	68/58	66/56	15	60	59	57	SE 3		WNW		
Barrow (S)	71 18	156 47	31	−45	−41	57/53	53/50	49/47	12	54	50	47	SW 8		SE		
Fairbanks AP (S)	64 49	147 52	436	−51	−47	82/62	78/60	75/59	24	64	62	60	N 5		S		
Juneau AP	58 22	134 35	12	−4	1	74/60	70/58	67/57	15	61	59	58	N 7		W		
Kodiak	57 45	152 29	73	10	13	69/58	65/56	62/55	10	60	58	56	WNW 14		NW		
Nome AP	64 30	165 26	13	−31	−27	66/57	62/55	59/54	10	58	56	55	N 4		W		
ARIZONA																	
Douglas AP	31 27	109 36	4098	27	31	98/63	95/63	93/63	31	70	69	68				104.4	14.0
Flagstaff AP	35 08	111 40	7006	−2	4	84/55	82/55	80/54	31	61	60	59	NE 5		SW	90.0	−11.6
Fort Huachuca AP (S)	31 35	110 20	4664	24	28	95/62	92/62	90/62	27	69	68	67	SW 5		W		
Kingman AP	35 12	114 01	3539	18	25	103/65	100/64	97/64	30	70	69	69					
Nogales	31 21	110 55	3800	28	32	99/64	96/64	94/64	31	71	70	69	SW 5		W		
Phoenix AP (S)	33 26	112 01	1112	31	34	109/71	107/71	105/71	27	76	75	75	E 4		W	112.8	26.7
Prescott AP	34 39	112 26	5010	4	9	96/61	94/60	92/60	30	66	65	64					
Tucson AP (S)	32 07	110 56	2558	28	32	104/66	102/66	100/66	26	72	71	71	SE 6		WNW	108.9	.3
Winslow AP	35 01	110 44	4895	5	10	97/61	95/60	93/60	32	66	65	64	SW 6		WSW	102.7	−.4
Yuma AP	32 39	114 37	213	36	39	111/72	109/72	107/71	27	79	78	77	NNE 6		WSW	114.8	30.8
ARKANSAS																	
Blytheville AFB	35 57	89 57	264	10	15	96/78	94/77	91/76	21	81	80	78	N 8		SSW		
Camden	33 36	92 49	116	18	23	98/76	96/76	94/76	21	80	79	78				101.0	13.9
El Dorado AP	33 13	92 49	277	18	23	98/76	96/76	94/76	21	80	79	78	S 6		SE	99.4	−.3
Fayetteville AP	36 00	94 10	1251	7	12	97/72	94/73	92/73	23	77	76	75	NE 9		SSW		
Fort Smith AP	35 20	94 22	463	12	17	101/75	98/76	95/76	24	80	79	78	NW 8		SW	101.9	7.0
Hot Springs	34 29	93 06	535	17	23	101/77	97/77	94/77	22	80	79	78	N 8		SW	103.0	10.6
Jonesboro	35 50	90 42	345	10	15	98/76	94/77	91/76	21	81	80	78				101.7	7.3
Little Rock AP (S)	34 44	92 14	257	15	20	99/76	96/77	94/77	22	80	79	78	N 9		SSW	99.0	11.2
Pine Bluff AP	34 18	92 05	241	16	22	100/78	97/77	95/78	22	81	80	80	N 7		SW	102.2	13.1
Texarkana AP	33 27	93 59	389	18	23	98/76	96/77	93/76	21	80	79	78	WNW 9		SSW	104.8	14.0

[a] AP or AFB following the station name designates airport or Airforce base temperature observations. Co designates office locations within an urban area that are affected by the surrounding area. Undesignated stations are semirural and may be compared to airport data.

[b] Winter design data are based on the 3-month period, December through February.
[c] Summer design data are based on the 4-month period, June through September.
[d] Mean wind speeds occurring coincidentally with the 99.5% dry-bulb winter design temperature.

Table 3-1. Design temperature criteria. (Copyright 1989 by the American Society of Heating, Refrigerating and Air-Conditioning Engineers, Inc., from the ASHRAE *Handbook—Fundamentals.* Used by permission.)

Col. 1	Col. 2		Col. 3		Col. 4	Col. 5 Winter,[b] °F Design Dry-Bulb		Col. 6 Summer,[c] °F Design Dry-Bulb and Coincident Wet-Bulb			Col. 7 Mean Daily	Col. 8 Design Wet-Bulb			Col. 9 Prevailing Wind				Col. 10 Temp. °F Median of Annual Extr.	
State and Station[a]	Lat.		Long.		Elev.			Mean	Coincident		Range				Winter		Summer			
	°	′	°	′	Feet	99%	97.5%	1%	2.5%	5%		1%	2.5%	5%	Knots[d]				Max.	Min.
CALIFORNIA																				
Bakersfield AP	35	25	119	03	475	30	32	104/70	101/69	98/68	32	73	71	70	ENE	5	WNW		109.8	25.3
Barstow AP	34	51	116	47	1927	26	29	106/68	104/68	102/67	37	73	71	70	WNW	7	W		110.4	17.4
Blythe AP	33	37	114	43	395	30	33	112/71	110/71	108/70	28	75	75	74					116.8	24.1
Burbank AP	34	12	118	21	775	37	39	95/68	91/68	88/67	25	71	70	69	NW	3	S			
Chico	39	48	121	51	238	28	30	103/69	101/68	98/67	36	71	70	68	NW	5	SSE		109.0	22.6
Concord	37	58	121	59	200	24	27	100/69	97/68	94/67	32	71	70	68	WNW	5	NW			
Covina	34	05	117	52	575	32	35	98/69	95/68	92/67	31	73	71	70						
Crescent City AP	41	46	124	12	40	31	33	68/60	65/59	63/58	18	62	60	59						
Downey	33	56	118	08	116	37	40	93/70	89/70	86/69	22	72	71	70						
El Cajon	32	49	116	58	367	42	44	83/69	80/69	78/68	30	71	70	68						
El Centro AP (S)	32	49	115	40	– 43	35	38	112/74	110/74	108/74	34	81	80	78	W	6	SE			
Escondido	33	07	117	05	660	39	41	89/68	85/68	82/68	30	71	70	69						
Eureka/Arcata AP	40	59	124	06	218	31	33	68/60	65/59	63/58	11	62	60	59	E	5	NW		75.8	29.7
Fairfield-Travis AFB	38	16	121	56	62	29	32	99/68	95/67	91/66	34	70	68	67	N	5	WSW			
Fresno AP (S)	36	46	119	43	328	28	30	102/70	100/69	97/68	34	72	71	70	E	4	WNW		108.7	25.8
Hamilton AFB	38	04	122	30	3	30	32	89/68	84/66	80/65	28	72	69	67	N	4	SE			
Laguna Beach	33	33	117	47	35	41	43	83/68	80/68	77/67	18	70	69	68						
Livermore	37	42	121	57	545	24	27	100/69	97/68	93/67	24	71	70	68	WNW	4	NW			
Lompoc, Vandenberg AFB	34	43	120	34	368	35	38	75/61	70/61	67/60	20	63	61	60	ESE	5	NW			
Long Beach AP	33	49	118	09	30	41	43	83/68	80/68	77/67	22	70	69	68	NW	4	WNW			
Los Angeles AP (S)	33	56	118	24	97	41	43	83/68	80/68	77/67	15	70	69	68	E	4	WSW			
Los Angeles Co (S)	34	03	118	14	270	37	40	93/70	89/70	86/69	20	72	71	70	NW	4	NW		98.1	35.9
Merced-castle AFB	37	23	120	34	188	29	31	102/70	99/69	96/68	36	72	71	70	ESE	4	NW			
Modesto	37	39	121	00	91	28	30	101/69	98/68	95/67	36	71	70	69					105.8	26.2
Monterey	36	36	121	54	39	35	38	75/63	71/61	68/61	20	64	62	61	SE	4	NW			
Napa	38	13	122	17	56	30	32	100/69	96/68	92/67	30	71	69	68					103.1	25.8
Needles AP	34	46	114	37	913	30	33	112/71	110/71	108/70	27	75	75	74					116.4	26.7
Oakland AP	37	49	122	19	5	34	36	85/64	80/63	75/62	19	66	64	63	E	5	WNW		93.0	31.8
Oceanside	33	14	117	25	26	41	43	83/68	80/68	77/67	13	70	69	68						
Ontario	34	03	117	36	952	31	33	102/70	99/69	96/67	36	74	72	71	E	4	WSW			
Oxnard	34	12	119	11	49	34	36	83/66	80/64	77/63	19	70	68	67						
Palmdale AP	34	38	118	06	2542	18	22	103/65	101/65	98/64	35	69	67	66	SW	5	WSW			
Palm Springs	33	49	116	32	411	33	35	112/71	110/70	108/70	35	76	74	73						
Pasadena	34	09	118	09	864	32	35	98/69	95/68	92/67	29	73	71	70					102.8	30.4
Petaluma	38	14	122	38	16	26	29	94/68	90/66	87/65	31	72	70	68					102.0	24.2
Pomona Co	34	03	117	45	934	28	30	102/70	99/69	95/68	36	74	72	71	E	4	W		105.7	26.2
Redding AP	40	31	122	18	495	29	31	105/68	102/67	100/66	32	71	69	68					109.2	26.0
Redlands	34	03	117	11	1318	31	33	102/70	99/69	96/68	33	74	72	71					106.7	27.1
Richmond	37	56	122	21	55	34	36	85/64	80/63	75/62	17	66	64	63						
Riverside-March AFB (S)	33	54	117	15	1532	29	32	100/68	98/68	95/67	37	72	71	70	N	4	NW		107.6	26.6
Sacramento AP	38	31	121	30	17	30	32	101/70	98/70	94/69	36	72	71	70	NNW	6	SW		105.1	27.6
Salinas AP	36	40	121	36	75	30	32	74/61	70/60	67/59	24	62	61	59						
San Bernardino, Norton AFB	34	08	117	13	1125	31	33	102/70	99/69	96/68	38	74	72	71	E	3	W		109.3	25.3
San Diego AP	32	44	117	10	13	42	44	83/69	80/69	78/68	12	71	70	68	NE	3	WNW		91.2	37.4
San Fernando	34	17	118	28	965	37	39	95/68	91/68	88/67	38	71	70	69						
San Francisco AP	37	37	122	23	8	35	38	82/64	77/63	73/62	20	65	64	62	S	5	NW			
San Francisco Co	37	46	122	26	72	38	40	74/63	71/62	69/61	14	64	62	61	W	5	W		91.3	35.9
San Jose AP	37	22	121	56	56	34	36	85/66	81/65	77/64	26	68	67	65	SE	4	NNW		98.6	28.2
San Luis Obispo	35	20	120	43	250	33	35	92/69	88/70	84/69	26	73	71	70	E	4	W		99.8	29.3
Santa Ana AP	33	45	117	52	115	37	39	89/69	85/68	82/68	28	71	70	69	E	3	SW		101.0	29.9
Santa Barbara MAP	34	26	119	50	10	34	36	81/67	77/66	75/65	24	68	67	66	NE	3	SW		97.1	31.7
Santa Cruz	36	59	122	01	125	35	38	75/63	71/61	68/61	28	64	62	61					97.5	26.8
Santa Maria AP (S)	34	54	120	27	236	31	33	81/64	76/63	73/62	23	65	64	63	E	4	WNW			
Santa Monica Co	34	01	118	29	64	41	43	83/68	80/68	77/67	16	70	69	68						
Santa Paula	34	21	119	05	263	33	35	90/68	86/67	84/66	36	71	69	68						
Santa Rosa	38	31	122	49	125	27	29	99/68	95/67	91/66	34	70	68	67	N	5	SE		102.5	23.4
Stockton AP	37	54	121	15	22	28	30	100/69	97/68	94/67	37	71	70	68	WNW	4	NW		104.1	24.5
Ukiah	39	09	123	12	623	27	29	99/69	95/68	91/67	40	70	68	67					108.1	21.6
Visalia	36	20	119	18	325	28	30	102/70	100/69	97/68	38	72	71	70					108.4	25.1
Yreka	41	43	122	38	2625	13	17	95/65	92/64	89/63	38	67	65	64					102.8	7.1
Yuba City	39	08	121	36	80	29	31	104/68	101/67	99/66	36	71	69	68						
COLORADO																				
Alamosa AP	37	27	105	52	7537	– 21	– 16	84/57	82/57	80/57	35	62	61	60						
Boulder	40	00	105	16	5445	2	8	93/59	91/59	89/59	27	64	63	62					96.0	– 8.4
Colorado Springs AP	38	49	104	43	6145	– 3	2	91/58	88/57	86/57	30	63	62	61	N	9	S		92.3	– 12.1
Denver AP	39	45	104	52	5283	– 5	1	93/59	91/59	89/59	28	64	63	62	S	8	SE		96.8	– 10.4
Durango	37	17	107	53	6550	– 1	4	89/59	87/59	85/59	30	64	63	62					92.4	– 11.2
Fort Collins	40	35	105	05	4999	– 10	– 4	93/59	91/59	89/59	28	64	63	62					95.2	– 18.1
Grand Junction AP (S)	39	07	108	32	4843	2	7	96/59	94/59	92/59	29	64	63	62	ESE	5	WNW		99.9	– 3.4
Greeley	40	26	104	38	4648	– 11	– 5	96/60	94/60	92/60	29	65	64	63						
Lajunta AP	38	03	103	30	4160	– 3	2	100/68	98/68	95/67	31	72	70	69	W	8	S			
Leadville	39	15	106	18	10155	– 8	– 4	84/52	81/51	78/50	30	56	55	54					79.7	– 17.8
Pueblo AP	38	18	104	29	4641	– 7	0	97/61	95/61	92/61	31	67	66	65	W	5	SE		100.5	– 12.2
Sterling	40	37	103	12	3939	– 7	– 2	95/62	93/62	90/62	30	67	66	65					100.3	– 15.4
Trinidad AP	37	15	104	20	5740	– 2	3	93/61	91/61	89/61	32	66	65	64	W	7	WSW		96.8	– 10.5

Table 3-1. Design temperature criteria, continued. (Copyright 1989 by the American Society of Heating, Refrigerating and Air-Conditioning Engineers, Inc., from the ASHRAE *Handbook—Fundamentals*. Used by permission.)

State and Station[a]	Col. 2 Lat. ° '	Col. 3 Long. ° '	Col. 4 Elev. Feet	Winter,[b] °F Col. 5 Design Dry-Bulb 99%	97.5%	Summer,[c] °F Col. 6 Design Dry-Bulb and Mean Coincident Wet-Bulb 1%	2.5%	5%	Col. 7 Mean Daily Range	Col. 8 Design Wet-Bulb 1%	2.5%	5%	Prevailing Wind Col. 9 Winter Knots[d]	Summer	Temp. °F Col. 10 Median of Annual Extr. Max.	Min.
CONNECTICUT																
Bridgeport AP	41 11	73 11	25	6	9	86/73	84/71	81/70	18	75	74	73	NNW 13	WSW		
Hartford, Brainard Field	41 44	72 39	19	3	7	91/74	88/73	85/72	22	77	75	74	N 5	SSW	95.7	−4.4
New Haven AP	41 19	73 55	6	3	7	88/75	84/73	82/72	17	76	75	74	NNE 7	SW	93.0	−.2
New London	41 21	72 06	59	5	9	88/73	85/72	83/71	16	76	75	74				
Norwalk	41 07	73 25	37	6	9	86/73	84/71	81/70	19	75	74	73				
Norwich	41 32	72 04	20	3	7	89/75	86/73	83/72	18	76	75	74				
Waterbury	41 35	73 04	843	−4	2	88/83	85/71	82/70	21	75	74	72	N 8	SW		
Windsor Locks, Bradley Fld	41 56	72 41	169	0	4	91/74	88/72	85/71	22	76	75	73	N 8	SW		
DELAWARE																
Dover AFB	39 08	75 28	28	11	15	92/75	90/75	87/74	18	79	77	76	W 9	SW	97.0	7.0
Wilmington AP	39 40	75 36	74	10	14	92/74	89/74	87/73	20	77	76	75	WNW 9	WSW	95.4	4.9
DISTRICT OF COLUMBIA																
Andrews AFB	38 5	76 5	279	10	14	92/75	90/74	87/73	18	78	76	75				
Washington, National AP	38 51	77 02	14	14	17	93/75	91/74	89/74	18	78	77	76	WNW 11	S	97.6	7.4
FLORIDA																
Belle Glade	26 39	80 39	16	41	44	92/76	91/76	89/76	16	79	78	78			94.7	30.9
Cape Kennedy AP	28 29	80 34	16	35	38	90/78	88/78	87/78	15	80	79	79				
Daytona Beach AP	29 11	81 03	31	32	35	92/78	90/77	88/77	15	80	79	78	NW 8			
E Fort Lauderdale	26 04	80 09	10	42	46	92/78	91/78	90/78	15	80	79	79	NW 9	ESE		
Fort Myers AP	26 35	81 52	15	41	44	93/78	92/78	91/77	18	80	79	79	NNE 7	W	94.9	34.9
Fort Pierce	27 28	80 21	25	38	42	91/78	90/78	89/78	15	80	79	79			96.1	34.0
Gainesville AP (S)	29 41	82 16	152	28	31	95/77	93/77	92/77	18	80	79	78	W 6	W	97.8	23.3
Jacksonville AP	30 30	81 42	26	29	32	96/77	94/77	92/76	19	79	79	78	NW 7	SW	97.5	25.4
Key West AP	24 33	81 45	4	55	57	90/78	90/78	89/78	09	80	79	79	NNE 12	SE	92.0	51.5
Lakeland Co (S)	28 02	81 57	214	39	41	93/76	91/76	89/76	17	79	78	78	NNW 9	SSW		
Miami AP (S)	25 48	80 16	7	44	47	91/77	90/77	89/77	15	79	79	78	NNW 9	SE	92.5	39.0
Miami Beach Co	25 47	80 17	10	45	48	90/77	89/77	88/77	10	79	79	78				
Ocala	29 11	82 08	89	31	34	95/77	93/77	92/76	18	80	79	78			98.6	24.8
Orlando AP	28 33	81 23	100	35	38	94/76	93/76	91/76	17	79	78	78	NNW 9	SSW		
Panama City, Tyndall AFB	30 04	85 35	18	29	33	92/78	90/77	89/77	14	81	80	79	N 8	WSW		
Pensacola Co	30 25	87 13	56	25	29	94/77	93/77	91/77	14	80	79	79	NNE 7	SW	96.3	23.3
St. Augustine	29 58	81 20	10	31	35	92/78	89/78	87/78	16	80	79	79	NW 7	W	97.6	25.8
St. Petersburg	27 46	82 80	35	36	40	92/77	91/77	90/76	16	79	79	78	N 8	W	94.8	35.6
Sanford	28 46	81 17	89	35	38	94/76	93/76	91/76	17	79	78	78				
Sarasota	27 23	82 33	26	39	42	93/77	92/77	90/76	17	79	79	78				
Tallahassee AP (S)	30 23	84 22	55	27	30	94/77	92/76	90/76	19	80	79	78	NW 6	NW	97.6	20.9
Tampa AP (S)	27 58	82 32	19	36	40	92/77	91/77	90/76	17	79	79	78	N 8	W	95.0	31.5
West Palm Beach AP	26 41	80 06	15	41	45	92/78	91/78	90/78	16	80	79	79	NW 9	ESE		
GEORGIA																
Albany, Turner AFB	31 36	84 05	223	25	29	97/77	95/76	93/76	20	80	79	78	N 7	W	100.6	19.9
Americus	32 03	84 14	456	21	25	97/77	94/76	92/75	20	79	78	77			100.4	16.5
Athens	33 57	83 19	802	18	22	94/74	92/74	90/74	21	78	77	76	NW 9	WNW	98.7	13.5
Atlanta AP (S)	33 39	84 26	1010	17	22	94/74	92/74	90/73	19	77	76	75	NW 11	NW	95.7	11.9
Augusta AP	33 22	81 58	145	20	23	97/77	95/76	93/76	19	80	79	78	W 4	WSW	99.0	17.5
Brunswick	31 15	81 29	25	29	32	92/78	89/78	87/78	18	80	79	79			99.3	24.7
Columbus, Lawson AFB	32 31	84 56	242	21	24	95/76	93/76	91/75	21	79	78	77	NW 8	W		
Dalton	34 34	84 57	720	17	22	94/76	93/76	91/76	22	79	78	77				
Dublin	32 20	82 54	215	21	25	96/77	93/76	91/75	20	79	78	77			101.0	16.7
Gainsville	34 11	83 41	50	24	27	96/77	93/77	91/77	20	80	79	78	WNW 7	SW	98.7	21.9
Griffin	33 13	84 16	981	18	22	93/76	90/75	88/74	21	78	77	76			12	98
LaGrange	33 01	85 04	709	19	23	94/76	91/75	89/74	21	79	78	76			17	100
Macon AP	32 42	83 39	354	21	25	96/77	93/76	91/75	22	79	78	77	NW 8	WNW	17	100
Marietta, Dobbins AFB	33 55	84 31	1068	17	21	94/74	92/74	90/74	21	78	77	76	NNW 12	NW		
Savannah	32 08	81 12	50	24	27	96/77	93/77	91/77	20	80	79	78	WNW 7	SW	22	99
Valdosta-Moody AFB	30 58	83 12	233	28	31	96/77	94/77	92/76	20	80	79	78	WNW 6	W		
Waycross	31 15	82 24	148	26	29	96/77	94/77	91/76	20	80	79	78			100.0	19.5
HAWAII																
Hilo AP (S)	19 43	155 05	36	61	62	84/73	83/72	82/72	15	75	74	74	SW 6	NE		
Honolulu AP	21 20	157 55	13	62	63	87/73	86/73	85/72	12	76	75	74	ENE 12	ENE		
Kaneohe Bay MCAS	21 27	157 46	18	65	66	85/75	84/74	83/74	12	76	76	75	NNE 9	NE		
Wahiawa	21 03	158 02	900	58	59	86/73	85/72	84/72	14	75	74	73	WNW 5	E		
IDAHO																
Boise AP (S)	43 34	116 13	2838	3	10	96/65	94/64	91/64	31	68	66	65	SE 6	NW	103.2	.6
Burley	42 32	113 46	4156	−3	2	99/62	95/61	92/66	35	64	63	61			98.6	−8.3
Coeur D'Alene AP	47 46	116 49	2972	−8	−1	89/62	86/61	83/60	31	64	63	61			99.9	−4.5
Idaho Falls AP	43 31	112 04	4741	−11	−6	89/61	87/61	84/59	38	65	63	61	N 9	S	96.2	−16.0
Lewiston AP	46 23	117 01	1413	−1	6	96/65	93/64	90/63	32	67	66	64	W 3	WNW	105.9	2.7
Moscow	46 44	116 58	2660	−7	0	90/63	87/62	84/61	32	65	64	62			98.0	−5.9
Mountain Home AFB	43 02	115 54	2996	6	12	99/64	97/63	94/62	36	66	65	63	ESE 7	NW	103.2	−6.5
Pocatello AP	42 55	112 36	4454	−8	−1	94/61	91/60	89/59	35	64	63	61	NE 5	W	97.9	−11.4
Twin Falls AP (S)	42 29	114 29	4150	−3	2	99/62	95/61	92/60	34	64	63	61	SE 6	NW	100.9	−5.1

Table 3-1. Design temperature criteria, continued. (Copyright 1989 by the American Society of Heating, Refrigerating and Air-Conditioning Engineers, Inc., from the ASHRAE *Handbook—Fundamentals*. Used by permission.)

Col. 1	Col. 2	Col. 3	Col. 4	Col. 5 Winter,[b] °F Design Dry-Bulb		Col. 6 Summer,[c] °F Design Dry-Bulb and Mean Coincident Wet-Bulb			Col. 7 Mean Daily	Col. 8 Design Wet-Bulb			Col. 9 Prevailing Wind		Col. 10 Temp. °F Median of Annual Extr.	
State and Station[a]	Lat. ° '	Long. ° '	Elev. Feet	99%	97.5%	1%	2.5%	5%	Range	1%	2.5%	5%	Winter Knots[d]	Summer	Max.	Min.
ILLINOIS																
Aurora	41 45	88 20	744	-6	-1	93/76	91/76	88/75	20	79	78	76			96.7	-13.0
Belleville, Scott AFB	38 33	89 51	453	1	6	94/76	92/76	89/75	21	79	78	76	WNW 8	S		
Bloomington	40 29	88 57	876	-6	-2	92/75	90/74	88/73	21	78	76	75			98.4	-9.6
Carbondale	37 47	89 15	417	2	7	95/77	93/77	90/76	21	80	79	77			100.9	-.8
Champaign/Urbana	40 02	88 17	777	-3	2	95/75	92/74	90/73	21	78	77	75				
Chicago, Midway AP	41 47	87 45	607	-5	0	94/74	91/73	88/72	20	77	75	74	NW 11	SW		
Chicago, O'Hare AP	41 59	87 54	658	-8	-4	91/74	89/74	86/72	20	77	76	74	WNW 9	SW		
Chicago Co	41 53	87 38	590	-3	2	94/75	91/74	88/73	15	79	77	75			96.1	-8.3
Danville	40 12	87 36	695	-4	1	93/75	90/74	88/73	21	78	77	75	W 10	SSW	98.2	-8.4
Decatur	39 50	88 52	679	-3	2	94/75	91/74	88/73	21	78	77	75	NW 10	SW	99.0	-8.1
Dixon	41 50	89 29	696	-7	-2	93/75	90/74	88/73	23	78	77	75			97.5	-13.5
Elgin	42 02	88 16	758	-7	-2	91/75	88/74	86/73	21	78	77	75				
Freeport	42 18	89 37	780	-9	-4	91/74	89/73	87/72	24	77	76	74				
Galesburg	40 56	90 26	764	-7	-2	93/75	91/75	88/74	22	78	77	75	WNW 8	SW		
Greenville	38 53	89 24	563	-1	4	94/76	92/75	89/74	21	79	78	76				
Joliet	41 31	88 10	582	-5	0	93/75	90/74	88/73	20	78	77	75	NW 11	SW		
Kankakee	41 05	87 55	625	-4	1	93/75	90/74	88/73	21	78	77	75				
La Salle/Peru	41 19	89 06	520	-7	-2	93/75	91/75	88/74	22	78	77	75				
Macomb	40 28	90 40	702	-5	0	95/76	92/76	89/75	22	79	78	76				
Moline AP	41 27	90 31	582	-9	-4	93/75	91/75	88/74	23	78	77	75	WNW 8	SW	96.8	-12.7
Mt Vernon	38 19	88 52	479	0	5	95/76	92/75	89/74	21	79	78	76			100.5	-2.9
Peoria AP	40 40	89 41	652	-8	-4	91/75	89/74	87/73	22	78	76	75	WNW 8	SW	98.0	-10.9
Quincy AP	39 57	91 12	769	-2	3	96/76	93/76	90/76	22	80	78	77	NW 11	SSW	101.1	-6.7
Rantoul, Chanute AFB	40 18	88 08	753	-4	1	94/75	91/74	89/73	21	78	77	75	W 10	SSW		
Rockford	42 21	89 03	741	-9	-4	91/74	89/73	87/72	24	77	76	74			97.4	-13.8
Springfield AP	39 50	89 40	588	-3	2	94/75	92/74	89/74	21	79	77	76	NW 10	SW	98.1	-7.2
Waukegan	42 21	87 53	700	-6	-3	92/76	89/74	87/73	21	78	76	75			96.5	-10.6
INDIANA																
Anderson	40 06	85 37	919	0	6	95/76	92/75	89/74	22	79	78	76	W 9	SW	95.1	-6.0
Bedford	38 51	86 30	670	0	5	95/76	92/75	89/74	22	79	78	76			97.5	-4.4
Bloomington	39 08	86 37	847	0	5	95/76	92/75	89/74	22	79	78	76	W 9	SW	97.8	-4.6
Columbus, Bakalar AFB	39 16	85 54	651	3	7	95/76	92/75	90/74	22	79	78	76	W 9	SW	98.3	-6.4
Crawfordsville	40 03	86 54	679	-2	3	94/75	91/74	88/73	22	79	77	76			98.4	-7.6
Evansville AP	38 03	87 32	381	4	9	95/76	93/75	91/75	22	79	78	77	NW 9	SW	98.2	.2
Fort Wayne AP	41 00	85 12	791	-4	1	92/73	89/72	87/72	24	77	75	74	WSW 10	SW		
Goshen AP	41 32	85 48	827	-3	1	91/73	89/73	86/72	23	77	75	74			96.8	-10.5
Hobart	41 32	87 15	600	-4	2	91/73	88/73	85/72	21	77	75	74			98.5	-8.5
Huntington	40 53	85 30	802	-4	1	92/73	89/72	87/72	23	77	75	74			96.9	-8.1
Indianapolis AP	39 44	86 17	792	-2	2	92/74	90/74	87/73	22	78	76	75	WNW 10	SW	96	-7
Jeffersonville	38 17	85 45	455	5	10	95/74	93/74	90/74	23	79	77	76			98	2
Kokomo	40 25	86 03	855	-4	0	91/74	90/73	88/73	22	77	75	74			98.2	-7.5
Lafayette	40 2	86 5	600	-3	3	94/74	91/73	88/73	22	78	76	75				
La Porte	41 36	86 43	810	-3	3	93/74	90/74	87/73	22	78	76	75			98.1	-10.5
Marion	40 29	85 41	859	-4	0	91/74	90/73	88/73	23	77	75	74			97.0	-8.6
Muncie	40 11	85 21	957	-3	2	92/74	90/73	87/73	22	76	76	75				
Peru, Grissom AFB	40 39	86 09	813	-6	-1	90/74	88/73	86/73	22	77	75	74	W 10	SW		
Richmond AP	39 46	84 50	1141	-2	2	92/74	90/74	87/73	22	78	76	75			94.8	-8.5
Shelbyville	39 31	85 47	750	-1	3	93/74	91/74	88/73	22	78	76	75			97.7	-6.0
South Bend AP	41 42	86 19	773	-3	1	91/73	89/73	86/72	22	77	75	74	SW 11	SSW	96.2	-9.2
Terre Haute AP	39 27	87 18	585	-2	4	95/75	92/74	89/73	22	79	77	76	NNW 7	SSW	98.3	-4.9
Valparaiso	41 31	87 02	801	-3	3	93/74	90/74	87/73	22	78	76	75			95.5	-11.0
Vincennes	38 41	87 32	420	1	6	95/75	92/74	90/73	22	79	77	76			100.3	-2.8
IOWA																
Ames (S)	42 02	93 48	1099	-11	-6	93/75	90/74	87/73	23	78	76	75			97.4	-17.8
Burlington AP	40 47	91 07	692	-7	-3	94/74	91/75	88/73	22	78	77	75	NW 9	SSW	98.6	-11.0
Cedar Rapids AP	41 53	91 42	863	-10	-5	91/76	88/75	86/74	23	78	77	75	NW 9	S	97.7	-15.6
Clinton	41 50	90 13	595	-8	-3	92/75	90/75	87/74	23	78	77	75			97.5	-13.8
Council Bluffs	41 20	95 49	1210	-8	-3	94/76	91/75	88/74	22	78	77	75				
Des Moines AP	41 32	93 39	938	-10	-5	94/75	91/74	88/73	23	78	77	75	NW 11	S	98.2	-14.2
Dubuque	42 24	90 42	1056	-12	-7	90/74	88/73	86/72	22	77	75	74	N 10	SSW	95.2	-15.0
Fort Dodge	42 33	94 11	1162	-12	-7	91/74	88/74	86/72	23	77	75	74	NW 11	S	98.5	-19.1
Iowa City	41 38	91 33	661	-11	-6	92/76	89/76	87/74	22	80	78	76	NW 9	SSW	97.4	-15.2
Keokuk	40 24	91 24	574	-5	0	95/75	92/75	89/74	22	79	77	76			98.4	-8.8
Marshalltown	42 04	92 56	898	-12	-7	92/76	90/75	88/73	23	78	77	75			98.5	-13.4
Mason City AP	43 09	93 20	1213	-15	-11	90/74	88/73	85/72	24	77	75	74	NW 11	S	96.5	-21.7
Newton	41 41	93 02	936	-10	-5	94/75	91/74	88/73	23	78	77	75			98.2	-14.7
Ottumwa AP	41 06	92 27	840	-8	-4	94/75	91/74	88/73	22	78	77	75			99.1	-12.0
Sioux City AP	42 24	96 23	1095	-11	-7	95/74	92/74	89/73	24	78	77	75	NNW 9	S	99.9	-17.7
Waterloo	42 33	92 24	868	-15	-10	91/76	89/75	86/74	23	78	77	75	NW 9	S	97.7	-19.8

Table 3-1. Design temperature criteria, continued. (Copyright 1989 by the American Society of Heating, Refrigerating and Air-Conditioning Engineers, Inc., from the ASHRAE *Handbook—Fundamentals*. Used by permission.)

State and Station	Lat. ° '	Long. ° '	Elev. Feet	Winter, °F Design Dry-Bulb 99%	97.5%	Summer, °F Design Dry-Bulb and Mean Coincident Wet-Bulb 1%	2.5%	5%	Mean Daily Range	Design Wet-Bulb 1%	2.5%	5%	Prevailing Wind Winter Knots	Summer Knots	Temp. °F Median of Annual Extr. Max.	Min.
KANSAS																
Atchison	39 34	95 07	945	− 2	2	96/77	93/76	91/76	23	81	79	77			100.5	− 8.8
Chanute AP	37 40	95 29	981	3	7	100/74	97/74	94/74	23	78	77	76	NNW 11	SSW	102.8	− 2.8
Dodge City AP (S)	37 46	99 58	2582	0	5	100/69	97/69	95/69	25	74	73	71	N 12	SSW	102.9	− 7.0
El Dorado	37 49	96 50	1282	3	7	101/72	98/73	96/73	24	77	76	75			103.5	− 5.0
Emporia	38 20	96 12	1210	1	5	100/74	97/74	94/73	25	78	77	76			102.4	− 6.4
Garden City AP	37 56	100 44	2880	− 1	4	99/69	96/69	94/69	28	74	73	71				
Goodland AP	39 22	101 42	3654	− 5	0	99/66	96/65	93/66	31	71	70	68	WSW 10	S	103.2	− 10.4
Great Bend	38 21	98 52	1889	0	4	101/73	98/73	95/73	28	78	76	75				
Hutchinson AP	38 04	97 52	1542	4	8	102/72	99/72	97/72	28	77	75	74	N 14	S	105.3	− 6.1
Liberal	37 03	100 58	2870	2	7	99/68	96/68	94/68	28	73	72	71			105.8	− 3.8
Manhattan, Ft Riley (S)	39 03	96 46	1065	− 1	3	99/75	95/75	92/74	24	78	77	76	NNE 8	S	104.5	− 8.6
Parsons	37 20	95 31	899	5	9	100/74	97/74	94/74	23	79	77	76	NNW 11	SSW		
Russell AP	38 52	98 49	1866	0	4	101/73	98/73	95/73	29	78	76	75				
Salina	38 48	97 39	1272	0	5	103/74	100/74	97/73	26	78	77	75	N 8	SSW		
Topeka AP	39 04	95 38	877	0	4	99/75	96/75	93/74	24	79	78	76	NNW 10	S	101.8	− 6.4
Wichita AP	37 39	97 25	1321	3	7	101/72	98/73	96/73	23	77	76	75	NNW 12	SSW	102.5	− 2.8
KENTUCKY																
Ashland	38 33	82 44	546	5	10	94/76	91/74	89/73	22	78	77	75	W 6	SW	97.4	.8
Bowling Green AP	35 58	86 28	535	4	10	94/77	92/75	89/74	21	79	77	76			99.9	1.2
Corbin AP	36 57	84 06	1174	4	9	94/73	92/73	89/72	23	77	76	75				
Covington AP	39 03	84 40	869	1	6	92/73	90/72	88/72	22	77	75	74	W 9	SW		
Hopkinsville, Ft Campbell	36 40	87 29	571	4	10	94/77	92/75	89/74	21	79	77	76	N 6	W	100.1	− .4
Lexington AP (S)	38 02	84 36	966	3	8	93/73	91/73	88/72	22	77	76	75	WNW 9	SW	95.3	− .5
Louisville AP	38 11	85 44	477	5	10	95/74	93/74	90/74	23	79	77	76	NW 8	SW	97.4	1.2
Madisonville	37 19	87 29	439	5	10	96/76	93/75	90/75	22	79	78	77				
Owensboro	37 45	87 10	407	5	10	97/76	94/75	91/75	23	79	78	77	NW 9	SW	98.0	− .2
Paducah AP	37 04	88 46	413	7	12	98/76	95/75	92/75	20	79	78	77				
LOUISIANA																
Alexandria AP	31 24	92 18	92	23	27	95/77	94/77	92/77	20	80	79	78	N 7	S	100.1	− 5.7
Baton Rouge AP	30 32	91 09	64	25	29	95/77	93/77	92/77	19	80	80	79	ENE 8	W	98.0	21.4
Bogalusa	30 47	89 52	103	24	28	95/77	93/77	92/77	19	80	80	79			99.3	20.2
Houma	29 31	90 40	13	31	35	93/78	93/78	92/77	15	81	80	79			97.2	22.5
Lafayette AP	30 12	92 00	42	26	30	95/78	94/78	92/78	18	81	80	79	N 8	SW	98.2	22.6
Lake Charles AP (S)	30 07	93 13	9	27	31	95/77	93/77	92/77	17	80	79	79	N 9	SSW	99.2	20.5
Minden	32 36	93 18	250	20	25	99/77	96/76	94/76	20	79	79	78			101.7	− 4.9
Monroe AP	32 31	92 02	79	20	25	99/77	96/76	94/76	20	79	79	78	N 9	S	101.1	− 5.9
Natchitoches	31 46	93 05	130	22	26	97/77	95/77	93/77	20	80	79	78				
New Orleans AP	29 59	90 15	4	29	33	93/78	92/78	90/77	16	81	80	79	NNE 9	SSW	96.3	27.7
Shreveport AP (S)	32 28	93 49	254	20	25	99/77	96/76	94/76	20	79	79	78	N 9	S		
MAINE																
Augusta AP	44 19	69 48	353	− 7	− 3	88/73	85/70	82/68	22	74	72	70	NNE 10	WNW		
Bangor, Dow AFB	44 48	68 50	192	− 11	− 6	86/70	83/68	80/67	22	73	71	69	WNW 7	S		
Caribou AP (S)	46 52	68 01	624	− 18	− 13	84/69	81/67	78/66	21	71	69	67	WSW 10	SW		
Lewiston	44 02	70 15	200	− 7	− 2	88/73	85/70	82/68	22	74	72	70			94.0	− 13.7
Millinocket AP	45 39	68 42	413	− 13	− 9	87/69	83/68	80/66	22	72	70	68	WNW 11	WNW	92.4	− 23.0
Portland (S)	43 39	70 19	43	− 6	− 1	87/72	84/71	81/69	22	74	72	70	W 7	S	93.5	− 9.9
Waterville	44 32	69 40	302	− 8	− 4	87/72	84/69	81/68	22	74	72	70				
MARYLAND																
Baltimore AP	39 11	76 40	148	10	13	94/75	91/75	89/74	21	78	77	76	W 9	WSW		
Baltimore Co	39 20	76 25	20	14	17	92/77	89/76	87/75	17	80	78	76	WNW 9	S	97.9	7.2
Cumberland	39 37	78 46	790	6	10	92/75	89/74	87/74	22	77	76	75	WNW 10	W		
Frederick AP	39 27	77 25	313	8	12	94/76	91/75	88/74	22	78	77	76	N 9	WNW		
Hagerstown	39 42	77 44	704	8	12	94/75	91/74	89/74	22	77	76	75	WNW 10	W		
Salisbury (S)	38 20	75 30	59	12	16	93/75	91/75	88/74	18	79	77	76			96.8	7.4
MASSACHUSETTS																
Boston AP (S)	42 22	71 02	15	6	9	91/73	88/71	85/70	16	75	74	72	WNW 16	SW	95.7	− 1.2
Clinton	42 24	71 41	398	− 2	2	90/72	87/71	84/69	17	75	73	72			91.7	− 8.5
Fall River	41 43	71 08	190	5	9	87/72	84/71	81/69	18	74	73	72	NW 10	SW	92.1	− 1.0
Framingham	42 17	71 25	170	3	6	89/72	86/71	83/69	17	74	73	71			96.0	− 7.7
Gloucester	42 35	70 41	10	2	5	89/73	86/71	83/70	15	75	74	72				
Greenfield	42 3	72 4	205	− 7	− 2	88/72	85/71	82/69	23	74	73	71				
Lawrence	42 42	71 10	57	− 6	0	90/73	87/72	84/70	22	76	74	73	NW 8	WSW	95.2	− 9.0
Lowell	42 39	71 19	90	− 4	1	91/73	88/72	85/70	21	76	74	73			95.1	− 8.5
New Bedford	41 41	70 58	79	5	9	85/72	82/71	80/69	19	74	73	72	NW 10	SW	91.4	2.2
Pittsfield AP	42 26	73 18	1194	− 8	− 3	87/71	84/70	81/68	23	73	72	70	NW 12	SW		
Springfield, Westover AFB	42 12	72 32	245	− 5	0	90/72	87/71	84/69	19	75	73	72	N 8	SSW	95.7	− 4.7
Taunton	41 54	71 04	20	5	9	89/73	86/72	83/70	18	75	74	73			92.9	− 9.8
Worcester AP	42 16	71 52	986	0	4	87/71	84/70	81/68	18	73	72	70	W 14	W		

Table 3-1. Design temperature criteria, continued. (Copyright 1989 by the American Society of Heating, Refrigerating and Air-Conditioning Engineers, Inc., from the ASHRAE *Handbook—Fundamentals*. Used by permission.)

State and Station[a]	Lat. ° '	Long. ° '	Elev. Feet	Winter °F Design Dry-Bulb 99%	97.5%	Summer °F Design Dry-Bulb and Coincident Wet-Bulb Mean 1%	2.5%	5%	Mean Daily Range	Design Wet-Bulb 1% 2.5% 5%	Prevailing Wind Winter Summer Knots[d]	Temp. °F Median of Annual Extr. Max.	Min.
MICHIGAN													
Adrian	41 55	84 01	754	− 1	3	91/73	88/72	85/71	23	76 75 73		97.2	− 7.0
Alpena AP	45 04	83 26	610	− 11	− 6	89/70	85/70	83/69	27	73 72 70	W 5 SW	93.9	− 14.8
Battle Creek AP	42 19	85 15	941	1	5	92/74	88/72	85/70	23	76 74 73	SW 8 SW		
Benton Harbor AP	42 08	86 26	643	1	5	91/72	88/72	85/70	20	75 74 72	SSW 8 WSW	95.1	− 2.6
Detroit	42 25	83 01	619	3	6	91/73	88/72	86/71	20	76 74 73	W 11 SW	88.8	− 16.1
Escanaba	45 44	87 05	607	− 11	− 7	87/70	83/69	80/68	17	73 71 69		95.3	− 9.9
Flint AP	42 58	83 44	771	− 4	1	90/73	87/72	85/71	25	76 74 72	SW 8 SW	95.4	− 5.6
Grand Rapids AP	42 53	85 31	784	1	5	91/72	88/72	85/70	24	75 74 72	WNW 8 WSW	94.1	− 6.8
Holland	42 42	86 06	678	2		88/72	86/71	83/70	22	75 73 72		96.5	− 7.8
Jackson AP	42 16	84 28	1020	1	5	92/74	88/72	85/70	23	76 74 73		95.9	− 6.7
Kalamazoo	42 17	85 36	955	1	5	92/74	88/72	85/70	23	76 74 73			
Lansing AP	42 47	84 36	873	− 3	1	90/73	87/72	84/70	24	75 74 72	SW 12 W	94.6	− 11.0
Marquette Co	46 34	87 24	735	− 12	− 8	84/70	81/69	77/66	18	72 70 68		94.5	− 11.8
Mt Pleasant	43 35	84 46	796	0	4	91/73	87/72	84/71	24	76 74 72		95.4	− 11.1
Muskegon AP	43 10	86 14	625	2	6	86/72	84/70	82/70	21	75 73 72	E 8 SW		
Pontiac	42 40	83 25	981	0	4	90/73	87/72	85/71	21	76 74 73		95.0	− 6.8
Port Huron	42 59	82 25	586	0	4	90/73	87/72	83/71	21	76 74 73	W 8 S		
Saginaw AP	43 32	84 05	667	0	4	91/73	87/72	84/71	23	76 74 72	WSW 7 SW	96.1	− 7.6
Sault Ste. Marie AP (S)	46 28	84 22	721	− 12	− 8	84/70	81/69	77/66	23	72 70 68	E 7 SW	89.8	− 21.0
Traverse City AP	44 45	85 35	624	− 3	1	89/72	86/71	83/69	22	75 73 71	SSW 9 SW	95.4	− 10.7
Ypsilanti	42 14	83 32	716	1	5	92/72	89/71	86/70	22	75 74 72	SW 10 SW		
MINNESOTA													
Albert Lea	43 39	93 21	1220	− 17	− 12	90/74	87/72	84/71	24	77 75 73			
Alexandria AP	45 52	95 23	1430	− 22	− 16	91/72	88/72	85/70	24	76 74 72		95.1	− 28.0
Bemidji AP	47 31	94 56	1389	− 31	− 26	88/69	85/69	81/67	24	73 71 69	N 8 S	94.5	− 36.9
Brainerd	46 24	94 08	1227	− 20	− 16	90/73	87/71	84/69	24	75 73 71			
Duluth AP	46 50	92 11	1428	− 21	− 16	85/70	82/68	79/66	22	72 70 68	WNW 12 WSW	90.9	− 27.4
Fairbault	44 18	93 16	940	17	− 12	91/74	88/72	85/71	24	77 75 73		95.8	− 24.3
Fergus Falls	46 16	96 04	1210	− 21	− 17	91/72	88/72	85/70	24	76 74 72		96.9	− 27.8
International Falls AP	48 34	93 23	1179	− 29	− 25	85/68	83/68	80/66	26	71 70 68	N 9 S	93.4	− 36.5
Mankato	44 09	93 59	1004	− 17	− 12	91/72	88/72	85/70	24	77 75 73			
Minneapolis/St. Paul AP	44 53	93 13	834	− 16	− 12	92/75	89/73	86/71	22	77 75 73	NW 8 S	96.5	− 22.0
Rochester AP	43 55	92 30	1297	− 17	− 12	90/74	87/72	84/71	24	77 75 73	NW 9 SSW		
St. Cloud AP (S)	45 35	94 11	1043	− 15	− 11	91/74	88/72	85/70	24	76 74 72		92.6	− 33.0
Virginia	47 30	92 33	1435	− 25	− 21	85/69	83/68	80/66	23	71 70 68		96.8	− 24.3
Willmar	45 07	95 05	1128	− 15	− 11	91/74	88/72	85/71	24	76 74 72			
Winona	44 03	91 38	652	− 14	− 10	91/75	88/73	85/72	24	77 75 74			
MISSISSIPPI													
Biloxi, Keesler AFB	30 25	88 55	26	28	31	94/79	92/79	90/78	16	82 81 80	N 8 S	23	98
Clarksdale	34 12	90 34	178	14	19	96/77	94/77	92/76	21	80 79 78		100.9	13.2
Columbus AFB	33 39	88 27	219	15	20	95/77	93/77	91/76	22	80 79 78	N 7 W	101.6	12.7
Greenville AFB	33 29	90 59	138	15	20	95/77	93/77	91/76	21	80 79 78		99.5	14.9
Greenwood	33 30	90 05	148	15	20	95/77	93/77	91/76	21	80 79 78		100.6	15.3
Hattiesburg	31 16	89 15	148	24	27	96/79	94/77	92/77	21	81 80 79		99.9	18.2
Jackson AP	32 19	90 05	310	21	25	97/76	95/76	93/76	21	79 78 78	NNW 6 NW	99.8	16.0
Laurel	31 40	89 10	236	24	27	96/78	94/77	92/77	21	81 80 79		99.7	17.8
Mccomb AP	31 15	90 28	469	21	26	96/77	94/76	92/76	18	80 79 78			
Meridian AP	32 20	88 45	290	19	23	97/77	95/76	93/76	22	80 79 78	N 6 WSW	98.3	15.7
Natchez	31 33	91 23	195	23	27	96/78	94/78	92/77	21	81 80 79		98.4	18.4
Tupelo	34 16	88 46	361	14	19	96/77	94/77	92/76	22	80 79 78		100.7	11.8
Vicksburg Co	32 24	90 47	262	22	26	97/78	95/78	93/77	21	81 80 79		96.9	18.0
MISSOURI													
Cape Girardeau	37 14	89 35	351	8	13	98/76	95/75	92/75	21	79 78 77		99.5	− 6.2
Columbia AP (S)	38 58	92 22	778	− 1	4	97/74	94/74	91/73	22	78 77 75	WNW 9 WSW	99.9	− 2.1
Farmington AP	37 46	90 24	928	3	8	96/76	93/75	90/76	22	80 78 77	NNW 11 SSW	98.4	− 7.6
Hannibal	39 42	91 21	489	− 2	3	96/76	93/75	90/76	22	78 77 76		101.2	− 6.1
Jefferson City	38 34	92 11	640	2	7	98/75	95/74	92/74	23	78 77 76			
Joplin AP	37 09	94 30	980	6	10	100/73	97/73	94/73	24	78 77 76	NNW 12 SSW		
Kansas City AP	39 07	94 35	791	2	6	99/75	96/74	93/74	20	78 77 76	NW 9 S	100.2	− 4.3
Kirksville AP	40 06	92 33	964	− 5	0	96/74	93/74	90/73	24	78 77 76		98.3	− 10.8
Mexico	39 11	91 54	775	− 1	4	97/74	94/74	91/73	22	78 77 76		101.2	− 8.0
Moberly	39 24	92 26	850	− 2	3	97/74	94/74	91/73	23	78 77 76			
Poplar Bluff	36 46	90 25	380	11	16	98/78	95/76	92/76	22	81 79 78		99.4	− 3.1
Rolla	37 59	91 43	1204	3	9	94/77	91/75	89/74	22	78 77 76		100.6	− 8.0
St. Joseph AP	39 46	94 55	825	− 3	2	96/77	93/76	91/76	23	81 79 77	NNW 9 S		
St. Louis AP	38 45	90 23	535	2	6	97/75	94/75	91/74	21	78 77 76	NW 9 WSW	99.1	− 2.7
St. Louis Co	38 39	90 38	462	3	8	98/75	94/75	91/74	18	78 77 76	NW 6 S		
Sikeston	36 53	89 36	325	9	15	98/77	95/76	92/75	21	80 78 77			
Sedalia, Whiteman AFB	38 43	93 33	869	− 1	4	95/76	92/76	90/75	22	79 78 76	NNW 7 SSW	100.0	− 5.1
Sikeston	36 53	89 36	325	9	15	98/77	95/76	92/75	21	80 78 77			
Springfield AP	37 14	93 23	1268	3	9	96/73	93/74	91/74	23	78 77 75	NNW 10 S	97.2	− 2.4

Table 3-1. Design temperature criteria, continued. (Copyright 1989 by the American Society of Heating, Refrigerating and Air-Conditioning Engineers, Inc., from the ASHRAE *Handbook—Fundamentals*. Used by permission.)

State and Station[a]	Lat. ° '	Long. ° '	Elev. Feet	Winter °F Design Dry-Bulb[b] 99%	97.5%	Summer °F Design Dry-Bulb and Mean Coincident Wet-Bulb[c] 1%	2.5%	5%	Mean Daily Range	Design Wet-Bulb 1%	2.5%	5%	Prevailing Wind Winter Knots[d]	Summer	Temp. °F Median of Annual Extr. Max.	Min.
MONTANA																
Billings AP	45 48	108 32	3567	−15	−10	94/64	91/64	88/63	31	67	66	64	NE 9	SW	100.5	−19.1
Bozeman	45 47	111 09	4448	−20	−14	90/61	87/60	84/59	32	63	62	60			92.2	−23.2
Butte AP	45 57	112 30	5553	−24	−17	86/58	83/56	80/56	35	60	58	57	S 5	NW	91.8	−26.3
Cut Bank AP	48 37	112 22	3838	−25	−20	88/61	85/61	82/60	35	64	62	61			94.7	−30.9
Glasgow AP (S)	48 25	106 32	2760	−22	−18	92/64	89/63	85/62	29	68	66	64	E 8	S		
Glendive	47 08	104 48	2476	−18	−13	95/66	92/64	89/62	29	69	67	65			103.3	−29.8
Great Falls AP (S)	47 29	111 22	3662	−21	−15	91/60	88/60	85/59	28	64	62	60	SW 7	WSW	98.0	−25.1
Havre	48 34	109 40	2492	−18	−11	94/65	90/64	87/63	33	68	66	65			99.7	−31.3
Helena AP	46 36	112 00	3828	−21	−16	91/60	88/60	85/59	32	64	62	61	N 12	WNW	95.6	−23.7
Kalispell AP	48 18	114 16	2974	−14	−7	91/62	87/61	84/60	34	65	63	62			94.4	−16.8
Lewiston AP	47 04	109 27	4122	−22	−16	90/62	87/61	83/60	30	65	63	62	NW 9	NW	96.2	−27.7
Livingstown AP	45 42	110 26	4618	−20	−14	90/61	87/60	84/59	32	63	62	60			97.2	−21.2
Miles City AP	46 26	105 52	2634	−20	−15	98/66	95/66	92/65	30	70	68	67	NW 7	SE	103.6	−27.7
Missoula AP	46 55	114 05	3190	−13	−6	92/62	88/61	85/60	36	65	63	62	ESE 7	NW	98.6	−13.9
NEBRASKA																
Beatrice	40 16	96 45	1235	−5	−2	99/75	95/74	92/74	24	78	77	76			103.1	−11.3
Chadron AP	42 50	103 05	3313	−8	−3	97/66	94/65	91/65	30	71	69	68				
Columbus	41 28	97 20	1450	−6	−2	98/74	95/73	92/73	25	77	76	75				
Fremont	41 26	96 29	1200	−6	−2	98/75	95/74	92/74	22	78	77	76				
Grand Island AP	40 59	98 19	1860	−8	−3	97/72	94/71	91/71	28	75	74	73	NNW 10	S	103.3	−14.2
Hastings	40 36	98 26	1954	−7	−3	97/72	94/71	91/71	27	75	74	73	NNW 10	S	103.5	−10.7
Kearney	40 44	99 01	2132	−9	−4	96/71	93/70	90/70	28	74	73	72			102.9	−13.7
Lincoln Co (S)	40 51	96 45	1180	−5	−2	99/75	95/74	92/74	24	78	77	76	N 8	S	102.0	−12.4
McCook	40 12	100 38	2768	−6	−2	98/69	95/69	91/69	28	74	72	71				
Norfolk	41 59	97 26	1551	−8	−4	97/74	93/74	90/73	30	78	77	75			102.0	−20.0
North Platte AP (S)	41 08	100 41	2779	−8	−4	97/69	94/69	90/69	28	74	72	71	NW 9	SSE	100.8	−15.8
Omaha AP	41 18	95 54	977	−8	−3	94/76	91/75	88/74	22	78	77	75	NW 8	S	100.2	−13.2
Scottsbluff AP	41 52	103 36	3958	−8	−3	95/65	92/65	90/64	31	70	68	67	NW 9	SE	101.6	−18.9
Sidney AP	41 13	103 06	4399	−8	−3	95/65	92/65	90/64	31	70	68	67				
NEVADA																
Carson City	39 10	119 46	4675	4	9	94/60	91/59	89/58	42	63	61	60	SSW 3	WNW	99.2	−5.0
Elko AP	40 50	115 47	5050	−8	−2	94/59	92/59	90/58	42	63	62	60	E 4	SW		
Ely AP (S)	39 17	114 51	6253	−10	−4	89/57	87/56	85/55	39	60	59	58	S 9	SSW		
Las Vegas AP (S)	36 05	115 10	2178	25	28	108/66	106/65	104/65	30	71	70	69	ENE 7	SW		
Lovelock AP	40 04	118 33	3903	8	12	98/63	96/63	93/62	42	66	65	64			103.0	−1.0
Reno AP (S)	39 30	119 47	4404	5	10	95/61	92/60	90/59	45	64	63	61	SSW 3	WNW		
Reno Co	39 30	119 47	4408	6	11	96/61	93/60	91/59	45	64	62	61			98.9	.2
Tonopah AP	38 04	117 05	5426	5	10	94/60	92/59	90/58	40	64	62	61	N 8	S		
Winnemucca AP	40 54	117 48	4301	−1	3	96/60	94/60	92/60	42	64	62	61	SE 10	W	100.1	−8.1
NEW HAMPSHIRE																
Berlin	44 03	71 01	1110	−14	−9	87/71	84/69	81/68	22	73	71	70			93.2	−24.7
Claremont	43 02	72 02	420	−9	−4	89/72	86/70	83/69	24	74	73	71				
Concord AP	43 12	71 30	342	−8	−3	90/72	87/70	84/69	26	74	73	71	NW 7	SW	94.8	−16.0
Keene	42 55	72 17	490	−12	−7	90/72	87/70	83/69	24	74	73	71			94.6	−18.9
Laconia	43 03	71 03	505	−10	−5	89/72	86/70	83/69	25	74	73	71				
Manchester, Grenier AFB	42 56	71 26	233	−8	−3	91/72	88/71	85/70	24	75	74	72	N 11	SW	93.7	−12.6
Portsmouth, Pease AFB	43 04	70 49	101	−2	2	89/73	85/71	83/70	22	75	74	72	W 8	W		
NEW JERSEY																
Atlantic City Co	39 23	74 26	11	10	13	92/74	89/74	86/72	18	78	77	75	NW 11	WSW	93.0	7.5
Long Branch	40 19	74 01	15	10	13	93/74	90/73	87/72	18	78	77	75			95.9	4.3
Newark AP	40 42	74 10	7	10	14	94/74	91/73	88/72	20	77	76	75	WNW 11	WSW		
New Brunswick	40 29	74 26	125	6	10	92/74	89/73	86/72	19	77	76	75				
Paterson	40 54	74 09	100	6	10	94/74	91/73	88/72	21	77	76	75				
Phillipsburg	40 41	75 11	180	1	6	92/73	89/72	86/71	21	76	75	74			97.4	−.7
Trenton Co	40 13	74 46	56	11	14	91/75	88/74	85/73	19	78	76	75	W 9	SW	96.2	4.2
Vineland	39 29	75 00	112	8	11	91/75	89/74	86/73	19	78	76	75				
NEW MEXICO																
Alamagordo, Holloman AFB	32 51	106 06	4093	14	19	98/64	96/64	94/64	30	69	68	67				
Albuquerque AP (S)	35 03	106 37	5311	12	16	96/61	94/61	92/61	27	66	65	64	N 7	W	98.1	5.1
Artesia	32 46	104 23	3320	13	19	103/67	100/67	97/67	30	72	71	70			105.5	3.7
Carlsbad AP	32 20	104 16	3293	13	19	103/67	100/67	97/67	28	72	71	70	N 6	SSE		
Clovis AP	34 23	103 19	4294	8	13	95/65	93/65	91/65	28	69	68	67			102.0	2.5
Farmington AP	36 44	108 14	5503	1	6	95/63	93/62	91/61	30	67	65	64	ENE 5	SW		
Gallup	35 31	108 47	6465	0	5	90/59	89/58	86/58	32	64	62	61				
Grants	35 10	107 54	6524	−1	4	89/59	88/58	85/57	32	64	62	61				
Hobbs AP	32 45	103 13	3690	13	18	101/66	99/66	97/66	29	71	70	69				
Las Cruces	32 18	106 55	4544	15	20	99/64	96/64	94/64	30	69	68	67	SE 5	SE		
Los Alamos	35 52	106 19	7410	5	9	89/60	87/60	85/60	32	62	61	60			89.8	−2.3
Raton AP	36 45	104 30	6373	−4	1	91/60	89/60	87/60	34	65	64	63				

Table 3-1. Design temperature criteria, continued. (Copyright 1989 by the American Society of Heating, Refrigerating and Air-Conditioning Engineers, Inc., from the ASHRAE *Handbook—Fundamentals*. Used by permission.)

Col. 1	Col. 2		Col. 3		Col. 4	Winter,[b] °F Col. 5		Summer,[c] °F Col. 6			Col. 7	Col. 8			Prevailing Wind Col. 9		Temp. °F Col. 10	
						Design Dry-Bulb		Design Dry-Bulb and Mean Coincident Wet-Bulb			Mean Daily	Design Wet-Bulb			Winter Summer		Median of Annual Extr.	
State and Station[a]	Lat.		Long.		Elev.	99%	97.5%	1%	2.5%	5%	Range	1%	2.5%	5%	Knots[d]		Max.	Min.
	°	'	°	'	Feet													
Roswell, Walker AFB	33	18	104	32	3676	13	18	100/66	98/66	96/66	33	71	70	69	N 6	SSE	103.0	2.7
Santa Fe Co	35	37	106	05	6307	6	10	90/61	88/61	86/61	28	63	62	61			90.1	– 1.2
Silver City AP	32	38	108	10	5442	5	10	95/61	94/60	91/60	30	66	64	63				
Socorro AP	34	03	106	53	4624	13	17	97/62	95/62	93/62	30	67	66	65				
Tucumcari AP	35	11	103	36	4039	8	13	99/66	97/66	95/65	28	70	69	68	NE 8	SW	102.7	1.1
NEW YORK																		
Albany AP (S)	42	45	73	48	275	– 6	– 1	91/73	88/72	85/70	23	75	74	72	WNW 8	S		
Albany Co	42	39	73	45	19	– 4	1	91/73	88/72	85/70	20	75	74	72			95.2	– 11.4
Auburn	42	54	76	32	715	– 3	2	90/73	87/71	84/70	22	75	73	72			92.4	– 9.5
Batavia	43	00	78	11	922	1	5	90/72	87/71	84/70	22	75	73	72			92.2	– 7.5
Binghamton AP	42	13	75	59	1590	– 2	1	86/71	83/69	81/68	20	73	72	70	WSW 10	WSW	92.9	– 9.3
Buffalo AP	42	56	78	44	705	2	6	88/71	85/70	83/69	21	74	73	72	W 10	SW	90.0	– 3.2
Cortland	42	36	76	11	1129	– 5	0	88/71	85/71	82/70	23	74	73	71			93.8	– 11.2
Dunkirk	42	29	79	16	692	4	9	88/73	85/72	83/71	18	75	74	72	SSW 10	WSW		
Elmira AP	42	10	76	54	955	– 4	1	89/73	86/71	83/70	24	74	73	71			96.2	– 6.7
Geneva (S)	42	45	76	54	613	– 3	2	90/73	87/71	84/70	22	75	73	72			96.1	– 6.5
Glens Falls	43	20	73	37	328	– 11	– 5	88/72	85/71	82/69	23	74	73	71	NNW 6	S		
Gloversville	43	02	74	21	760	– 8	– 2	89/72	86/71	83/69	23	75	74	72			93.2	– 14.6
Hornell	42	21	77	42	1325	– 4	0	88/71	85/70	82/69	24	74	73	72				
Ithaca (S)	42	27	76	29	928	– 5	0	88/71	85/71	82/70	24	74	73	71	W 6	SW		
Jamestown	42	07	79	14	1390	– 1	3	88/70	86/70	83/69	20	74	72	71	WSW 9	WSW		
Kingston	41	56	74	00	279	– 3	2	91/73	88/72	85/70	22	76	74	73				
Lockport	43	09	79	15	638	4	7	89/74	86/72	84/71	21	76	74	73	N 9	SW	92.2	– 4.8
Massena AP	44	56	74	51	207	– 13	– 8	86/70	83/69	80/68	20	73	72	70				
Newburgh, Stewart AFB	41	30	74	06	471	– 1	4	90/73	88/72	85/70	21	76	74	73	W 10	W		
NYC-Central Park (S)	40	47	73	58	157	11	15	92/74	89/73	87/72	17	76	75	74			94.9	3.8
NYC-Kennedy AP	40	39	3	47	13	12	15	90/73	87/72	84/71	16	76	75	74	WNW 4	SSW		
NYC-La Guardia AP	40	46	73	54	11	11	15	92/74	89/73	87/72	16	76	75	74	WNW 15	SW		
Niagara Falls AP	43	06	79	57	590	4	7	89/74	86/72	84/71	20	76	74	73	W 9	SW		
Olean	42	14	78	22	2119	– 2	2	87/71	84/71	81/70	23	74	73	71				
Oneonta	42	31	75	04	1775	– 7	– 4	86/71	83/69	80/68	24	73	72	70				
Oswego Co	43	28	76	33	300	1	7	86/73	83/71	80/70	20	75	73	72	E 7	WSW	91.3	– 7.4
Plattsburg AFB	44	39	73	28	235	– 13	– 8	86/70	83/69	80/68	22	73	72	70	NW 6	SE		
Poughkeepsie	41	38	73	55	165	0	6	92/74	89/74	86/72	21	77	75	74	NNE 6	SSW	98.1	– 5.6
Rochester AP	43	07	77	40	547	1	5	91/73	88/71	85/70	22	75	73	72	WSW 11	WSW		
Rome, Griffiss AFB	43	14	75	25	514	– 11	– 5	88/71	85/70	83/69	22	75	73	71	NW 5	W		
Schenectady (S)	42	51	73	57	377	– 4	1	90/73	87/72	84/70	22	75	74	72	WNW 8	S		
Suffolk County AFB	40	51	72	38	67	7	10	86/72	83/71	80/70	16	76	74	73	NW 9	SW		
Syracuse AP	43	07	76	07	410	– 3	2	90/73	87/71	84/70	20	75	73	72	N 7	WNW	93.	– 10.0
Utica	43	09	75	23	714	– 12	– 6	88/73	85/71	82/70	22	75	73	71	NW 12	W		
Watertown	43	59	76	01	325	– 11	– 6	86/73	83/71	81/70	20	75	73	72	E 7	WSW	91.7	– 19.6
NORTH CAROLINA																		
Asheville AP	35	26	82	32	2140	10	14	89/73	87/72	85/71	21	75	74	72	NNW 12	NNW	91.9	5.8
Charlotte AP	35	13	80	56	736	18	22	95/74	93/74	91/74	20	77	76	76	NNW 6	SW	97.8	12.6
Durham	35	52	78	47	434	16	20	94/75	92/75	90/75	20	78	77	76			98.9	9.6
Elizabeth City AP	36	16	76	11	12	12	19	93/78	91/77	89/76	18	80	78	78	NW 8	SW		
Fayetteville, Pope AFB	35	10	79	01	218	17	20	95/76	92/76	90/75	20	79	78	77	N 6	SSW	99.1	13.1
Goldsboro, Seymour-Johnson	35	20	77	58	109	18	21	94/77	91/76	89/75	18	79	78	77	N 8	SW	99.8	13.0
Greensboro AP (S)	36	05	79	57	897	14	18	93/74	91/73	89/73	21	77	76	75	NE 7	SW	97.7	9.7
Greenville	35	37	77	25	75	18	21	93/77	91/76	89/75	19	79	78	77				
Henderson	36	22	78	25	480	12	15	95/77	92/76	90/76	20	79	78	77				
Hickory	35	45	81	23	1187	14	18	92/73	90/72	88/72	21	75	74	73			96.5	9.6
Jacksonville	34	50	77	37	95	20	24	92/78	90/78	88/77	18	80	79	78				
Lumberton	34	37	79	04	129	18	21	95/76	92/76	90/75	20	79	78	77				
New Bern AP	35	05	77	03	20	20	24	92/78	90/78	88/77	18	80	79	78			98.2	15.1
Raleigh/Durham AP (S)	35	52	78	47	434	16	20	94/75	92/75	90/75	20	78	77	76	N 7	SW	97.7	12.2
Rocky Mount	35	58	77	48	121	18	21	94/77	91/76	89/75	19	79	78	77				
Wilmington AP	34	16	77	55	28	23	26	93/79	91/78	89/77	18	81	80	79	N 8	SW	96.9	18.2
Winston-Salem AP	36	08	80	13	969	16	20	94/74	91/73	89/73	20	76	75	74	NW 8	WSW		
NORTH DAKOTA																		
Bismarck AP (S)	46	46	100	45	1647	– 23	– 19	95/68	91/68	88/67	27	73	71	70	WNW 7	S	100.3	– 31.5
Devils Lake	48	07	98	54	1450	– 25	– 21	91/69	88/68	85/66	25	73	71	69			97.5	– 30.4
Dickinson AP	46	48	102	48	2585	– 21	– 17	94/68	90/66	87/65	25	71	69	68	WNW 12	SSE	101.0	– 31.3
Fargo AP	46	54	96	48	896	– 22	– 18	92/73	89/71	85/69	25	76	74	72	SSE 11	S	97.3	– 29.7
Grand Forks AP	47	57	97	24	911	– 26	– 22	91/70	87/70	84/68	25	74	72	70	N 8	S	97.6	– 29.0
Jamestown AP	46	55	98	41	1492	– 22	– 18	94/70	90/69	87/68	26	74	74	71			101.3	– 27.9
Minot AP	48	25	101	21	1668	– 24	– 20	92/68	89/67	86/65	25	72	70	68	WSW 10	S		
Williston	48	09	103	35	1876	– 25	– 21	91/68	88/67	85/65	25	72	70	68			99.7	– 32.9

Table 3-1. Design temperature criteria, continued. (Copyright 1989 by the American Society of Heating, Refrigerating and Air-Conditioning Engineers, Inc., from the ASHRAE *Handbook—Fundamentals.* Used by permission.)

Col. 1	Col. 2		Col. 3		Col. 4	Col. 5 Winter,[b] °F		Col. 6 Summer,[c] °F			Col. 7	Col. 8			Col. 9 Prevailing Wind		Col. 10 Temp. °F	
						Design Dry-Bulb		Design Dry-Bulb and Mean Coincident Wet-Bulb			Mean Daily	Design Wet-Bulb			Winter	Summer	Median of Annual Extr.	
State and Station[a]	Lat.		Long.		Elev.	99%	97.5%	1%	2.5%	5%	Range	1%	2.5%	5%	Knots[d]		Max.	Min.
	°	'	°	'	Feet													
OHIO																		
Akron-Canton AP	40	55	81	26	1208	1	6	89/72	86/71	84/70	21	75	73	72	SW 9	SW	94.4	-4.6
Ashtabula	41	51	80	48	690	4	9	88/73	85/72	83/71	18	75	74	72				
Athens	39	20	82	06	700	0	6	95/75	92/74	90/73	22	78	76	74				
Bowling Green	41	23	83	38	675	-2	2	92/73	89/73	86/71	23	76	75	73			96.7	-7.3
Cambridge	40	04	81	35	807	1	7	93/75	90/74	87/73	23	78	76	75				
Chillicothe	39	21	83	00	640	0	6	95/75	92/74	90/73	22	78	76	74	W 8	WSW	98.2	-2.1
Cincinnati Co	39	09	84	31	758	1	6	92/73	90/72	88/72	21	77	75	74	W 9	SW	97.2	-.2
Cleveland AP (S)	41	24	81	51	777	1	5	91/73	88/72	86/71	22	76	74	73	SW 12	N	94.7	-3.1
Columbus AP (S)	40	00	82	53	812	0	5	92/73	90/73	87/72	24	77	75	74	W 8	SSW	96.0	-3.4
Dayton AP	39	54	84	13	1002	-1	4	91/73	89/72	86/71	20	76	75	73	WNW 11	SW	96.6	-4.5
Defiance	41	17	84	23	700	-1	4	94/74	91/73	88/72	24	77	76	74				
Findlay AP	41	01	83	40	804	2	3	92/74	90/73	87/72	24	77	76	74			97.4	-7.4
Fremont	41	20	83	07	600	-3	1	90/73	88/73	85/71	24	76	75	73				
Hamilton	39	24	84	35	650	0	5	92/73	90/72	87/71	22	76	75	73			98.2	-2.8
Lancaster	39	44	82	38	860	0	5	93/74	91/73	88/72	23	77	75	74				
Lima	40	42	84	02	975	-1	4	94/74	91/73	88/72	24	77	76	74	WNW 11	SW	96.0	-6.5
Mansfield AP	40	49	82	31	1295	0	5	90/73	87/72	85/72	22	76	74	73	W 8	SW	93.8	-10.7
Marion	40	36	83	10	920	0	5	93/74	91/73	88/72	23	77	76	74				
Middletown	39	31	84	25	635	0	5	92/73	90/72	87/71	22	76	75	73			95.8	-6.8
Newark	40	01	82	28	880	-1	5	94/73	92/73	89/72	23	77	75	74	W 8	SSW		
Norwalk	41	16	82	37	670	-3	1	90/73	88/73	85/71	22	76	75	73			97.3	-8.3
Portsmouth	38	45	82	55	540	5	10	95/76	92/74	89/73	22	78	77	75	W 8	SW	97.9	1.0
Sandusky Co	41	27	82	43	606	1	6	93/73	91/72	88/71	21	76	74	73			96.7	-1.9
Springfield	39	50	83	50	1052	-1	3	91/74	89/73	87/72	21	77	76	74	W 7	W		
Steubenville	40	23	80	38	992	1	5	89/72	86/71	84/70	22	74	73	72				
Toledo AP	41	36	83	48	669	-3	1	90/73	88/73	85/71	25	76	75	73	WSW 8	SW	95.4	-5.2
Warren	41	20	80	51	928	0	5	89/71	87/71	85/70	23	74	73	71				
Wooster	40	47	81	55	1020	1	6	89/72	86/71	84/70	22	75	73	72			94.0	-7.7
Youngstown AP	41	16	80	40	1178	-1	4	88/71	86/71	84/70	23	74	73	71	SW 10	SW		
Zanesville AP	39	57	81	54	900	1	7	93/75	90/74	87/73	23	78	76	75	W 6	WSW		
OKLAHOMA																		
Ada	34	47	96	41	1015	10	14	100/74	97/74	95/74	23	77	76	75				
Altus AFB	34	39	99	16	1378	11	16	102/73	100/73	98/73	25	77	76	75	N 10	S		
Ardmore	34	18	97	01	771	13	17	100/74	98/74	95/74	23	77	77	76				
Bartlesville	36	45	96	00	715	6	10	101/73	98/74	95/74	23	77	77	76				
Chickasha	35	03	97	55	1085	10	14	101/74	98/74	95/74	24	78	77	76				
Enid, Vance AFB	36	21	97	55	1307	9	13	103/74	100/74	97/74	24	79	77	76				
Lawton AP	34	34	98	25	1096	12	16	101/74	99/74	96/74	24	78	77	76				
McAlester	34	50	95	55	776	14	19	99/74	96/74	93/74	23	77	76	75	N 10	S		
Muskogee AP	35	40	95	22	610	10	15	101/74	98/75	95/75	23	79	78	77				
Norman	35	15	97	29	1181	9	13	99/74	96/74	94/74	24	77	76	75	N 10	S		
Oklahoma City AP (S)	35	24	97	36	1285	9	13	100/74	97/74	95/73	23	78	77	76	N 14	SSW		
Ponca City	36	44	97	06	997	5	9	100/74	97/74	94/74	24	77	76	76				
Seminole	35	14	96	40	865	11	15	99/74	96/74	94/74	23	77	76	75				
Stillwater (S)	36	10	97	05	984	8	13	100/74	96/74	93/74	24	77	76	75	N 12	SSW	103.7	1.6
Tulsa AP	36	12	95	54	650	8	13	101/74	98/75	95/75	22	79	78	77	N 11	SSW		
Woodward	36	36	99	31	2165	6	10	100/73	97/73	94/73	26	78	76	75			107.1	-1.3
OREGON																		
Albany	44	38	123	07	230	18	22	92/67	89/66	86/65	31	69	67	66			97.5	16.6
Astoria AP (S)	46	09	123	53	8	25	29	75/65	71/62	68/61	16	65	63	62	ESE 7	NNW		
Baker AP	44	50	117	49	3372	-1	6	92/63	89/61	86/60	30	65	63	61			97.5	-6.8
Bend	44	04	121	19	3595	-3	4	90/62	87/60	84/59	33	64	62	60			96.4	-5.8
Corvallis (S)	44	30	123	17	246	18	22	92/67	89/66	86/65	31	69	67	66	N 6	N	98.5	17.1
Eugene AP	44	07	123	13	359	17	22	92/67	89/66	86/65	31	69	67	66	N 7	N		
Grants Pass	42	26	123	19	925	20	24	99/69	96/68	93/67	33	71	69	68	N 5	N	103.6	16.4
Klamath Falls AP	42	09	121	44	4092	4	9	90/61	87/60	84/59	36	63	61	60	N 4	W	96.3	.9
Medford AP (S)	42	22	122	52	1298	19	23	98/68	94/67	91/66	35	70	68	67	S 4	WMW	103.8	15.0
Pendleton AP	45	41	118	51	1482	-2	5	97/65	93/64	90/62	29	66	65	63	NNW 6	WNW		
Portland AP	45	36	122	36	21	17	23	89/68	85/67	81/65	23	69	67	66	ESE 12	NW	96.6	18.3
Portland Co	45	32	122	40	75	18	24	90/68	86/67	82/65	21	69	67	66			97.6	20.5
Roseburg AP	43	14	123	22	525	18	23	93/67	90/66	87/65	30	69	67	66			99.6	19.5
Salem AP	44	55	123	01	196	18	23	92/68	88/66	84/65	31	69	68	66	N 6	N	98.9	15.9
The Dalles	45	36	121	12	100	13	19	93/69	89/68	85/66	28	70	68	67			105.1	7.9

Table 3-1. Design temperature criteria, continued. (Copyright 1989 by the American Society of Heating, Refrigerating and Air-Conditioning Engineers, Inc., from the ASHRAE *Handbook—Fundamentals*. Used by permission.)

Col. 1	Col. 2		Col. 3		Col. 4	Winter,[b] °F — Col. 5		Summer,[c] °F — Col. 6			Col. 7	Col. 8			Col. 9		Col. 10	
State and Station[a]	Lat.		Long.		Elev.	Design Dry-Bulb		Design Dry-Bulb and Coincident Wet-Bulb			Mean Daily	Design Wet-Bulb			Prevailing Wind		Median of Annual Extr.	
	°	′	°	′	Feet	99%	97.5%	Mean 1%	2.5%	5%	Range	1%	2.5%	5%	Winter	Summer	Max.	Min.
PENNSYLVANIA															Knots[d]			
Allentown AP	40	39	75	26	387	4	9	92/73	88/72	86/72	22	76	75	73	W 11	SW		
Altoona Co	40	18	78	19	1504	0	5	90/72	87/71	84/70	23	74	73	72	WNW 11	WSW	93.7	− 5.2
Butler	40	52	79	54	1100	1	6	90/73	87/72	85/71	22	75	74	73				
Chambersburg	39	56	77	38	640	4	8	93/75	90/74	87/73	23	77	76	75			97.4	− .3
Erie AP	42	05	80	11	731	4	9	88/73	85/72	83/71	18	75	74	72	SSW 0	WSW	91.3	− 2.2
Harrisburg AP	40	12	76	46	308	7	11	94/75	91/74	88/73	21	77	76	75	NW 11	WSW	96.5	3.7
Johnstown	40	19	78	50	2284	− 3	2	86/70	83/70	80/68	23	72	71	70	WNW 8	WSW	96.4	− 1.8
Lancaster	40	07	76	18	403	4	8	93/75	90/74	87/73	22	77	76	75	NW 11	WSW		
Meadville	41	38	80	10	1065	0	4	88/71	85/70	83/69	21	73	72	71			93.2	− 8.5
New Castle	41	01	80	22	825	2	7	91/73	88/72	86/71	23	75	74	73	WSW 10	WSW	94.7	− 6.4
Philadelphia AP	39	53	75	15	5	10	14	93/75	90/74	87/72	21	77	76	75	WNW 10	WSW	96.4	5.9
Pittsburgh AP	40	30	80	13	1137	1	5	89/72	86/71	84/70	22	74	73	72	WSW 10	WSW		
Pittsburgh Co	40	27	80	00	1017	3	7	91/72	88/71	86/70	19	74	73	72			94.6	− 1.1
Reading Co	40	20	75	38	266	9	13	92/73	89/72	86/72	19	76	75	73	W 11	SW	97.0	3.6
Scranton/Wilkes-Barre	41	20	75	44	930	1	5	90/72	87/71	84/70	19	74	73	72	SW 8	WSW	94.8	− 2.2
State College (S)	40	48	77	52	1175	3	7	90/72	87/71	84/70	23	74	73	72	NNW 8	WSW	93.2	− 3.6
Sunbury	40	53	76	46	446	2	7	92/73	89/72	86/70	22	75	74	73				
Uniontown	39	55	79	43	956	5	9	91/74	88/73	85/72	22	76	75	74			93.9	− 2.5
Warren	41	51	79	08	1280	− 2	4	89/71	86/71	83/70	24	74	73	72			93.3	− 10.7
West Chester	39	58	75	38	450	9	13	92/75	89/74	86/72	20	77	76	75				
Williamsport AP	41	15	76	55	524	2	7	92/73	89/72	86/70	23	75	74	73	W 9	WSW	95.5	− 4.6
York	39	55	76	45	390	8	12	94/75	91/74	88/73	22	77	76	75			97.0	− 2.4
RHODE ISLAND																		
Newport (S)	41	30	71	20	10	5	9	88/73	85/72	82/70	16	76	75	73	WNW 10	SW		
Providence AP	41	44	71	26	51	5	9	89/73	86/72	83/70	19	75	74	73	WNW 11	SW	94.6	− .5
SOUTH CAROLINA																		
Anderson	34	30	82	43	774	19	23	94/74	92/74	90/74	21	77	76	75			99.5	13.3
Charleston AFB (S)	32	54	80	02	45	24	27	93/78	91/78	89/77	18	81	80	79	NNE 8	SW		
Charleston Co	32	54	79	58	3	25	28	94/78	92/78	90/77	13	81	80	79			97.8	21.4
Columbia AP	33	57	81	07	213	20	24	97/76	95/75	93/75	22	79	78	77	W 6	SW	100.6	16.2
Florence AP	34	11	79	43	147	22	25	94/77	92/77	90/76	21	80	79	78	N 7	SW	99.5	16.5
Georgetown	33	23	79	17	14	23	26	92/79	90/78	88/77	18	81	80	79	N 7	SSW	98.2	19.1
Greenville AP	34	54	82	13	957	18	22	93/74	91/74	89/74	21	77	76	75	NW 8	SW	97.3	12.6
Greenwood	34	10	82	07	620	18	22	95/74	93/74	91/74	21	78	77	76			99.5	14.1
Orangeburg	33	30	80	52	260	20	24	97/76	95/75	93/75	20	79	78	77			101.2	18.0
Rock Hill	34	59	80	58	470	19	23	96/75	94/74	92/74	20	78	77	76				
Spartanburg AP	34	58	82	00	823	18	22	93/74	91/74	89/74	20	77	76	75			99.5	13.9
Sumter, Shaw AFB	33	54	80	22	169	22	25	95/77	92/76	90/75	21	79	78	77	NNE 6	W	100.0	15.4
SOUTH DAKOTA																		
Aberdeen AP	45	27	98	26	1296	− 19	− 15	94/73	91/72	88/70	27	77	75	73	NNW 8	S	102.3	− 28.1
Brookings	44	18	96	48	1637	− 17	− 13	95/73	92/72	89/71	25	77	75	73			97.8	− 26.5
Huron AP	44	23	98	13	1281	− 18	− 14	96/73	93/72	90/71	28	77	75	73	NNW 8	S	101.5	− 25.8
Mitchell	43	41	98	01	1346	− 15	− 10	96/72	93/71	90/70	28	76	75	73			103.0	− 22.7
Pierre AP	44	23	100	17	1742	− 15	− 10	99/71	95/71	92/69	29	75	74	72	NW 11	SSE	105.7	− 20.6
Rapid City AP (S)	44	03	103	04	3162	− 11	− 7	95/66	92/65	89/65	28	71	69	67	NNW 10	SSE	100.9	− 19.0
Sioux Falls AP	43	34	96	44	1418	− 15	− 11	94/73	91/72	88/71	24	76	75	73	NW 8	S		
Watertown AP	44	55	97	09	1738	− 19	− 15	94/73	91/72	88/71	26	76	75	73			97.8	− 26.5
Yankton	42	55	97	23	1302	− 13	− 7	94/73	91/72	88/71	25	77	76	74			100.8	− 19.1
TENNESSEE																		
Athens	35	26	84	35	940	13	18	95/74	92/73	90/73	22	77	76	75				
Bristol-Tri City AP	36	29	82	24	1507	9	14	91/72	89/72	87/71	22	75	75	73	WNW 6	SW		
Chattanooga AP	35	02	85	12	665	13	18	96/75	93/74	91/74	22	78	77	76	NNW 8	WSW	97.2	9.8
Clarksville	36	33	87	22	382	6	12	95/76	93/74	90/74	21	78	77	76			99.8	3.7
Columbia	35	38	87	02	690	10	15	97/75	94/74	91/74	21	78	77	76				
Dyersburg	36	01	89	24	344	10	15	96/78	94/77	91/76	21	81	80	78				
Greenville	36	04	82	50	1319	11	16	92/73	90/72	88/72	22	76	75	74			95.6	.8
Jackson AP	35	36	88	55	423	11	16	98/76	95/75	92/75	21	79	78	77			99.2	6.6
Knoxville AP	35	49	83	59	980	13	19	94/74	92/73	90/73	21	77	76	75	NE 8	W	96.0	7.0
Memphis AP	35	03	90	00	258	13	18	98/77	95/76	93/76	21	80	79	78	N 10	SW	97.9	10.4
Murfreesboro	34	55	86	28	600	9	14	97/75	94/74	91/74	22	78	77	76			97.7	`4.5
Nashville AP (S)	36	07	86	41	590	9	14	97/75	94/74	91/74	21	78	77	76	NW 8	WSW		
Tullahoma	35	23	86	05	1067	8	13	96/74	93/73	91/73	22	77	76	75	NW 9	WSW	96.7	3.7

Table 3-1. Design temperature criteria, continued. (Copyright 1989 by the American Society of Heating, Refrigerating and Air-Conditioning Engineers, Inc., from the ASHRAE *Handbook—Fundamentals*. Used by permission.)

Col. 1	Col. 2 Lat. ° '	Col. 3 Long. ° '	Col. 4 Elev. Feet	Col. 5 Winter, °F Design Dry-Bulb 99%	97.5%	Col. 6 Summer, °F Design Dry-Bulb and Mean Coincident Wet-Bulb 1%	2.5%	5%	Col. 7 Mean Daily Range	Col. 8 Design Wet-Bulb 1%	2.5%	5%	Col. 9 Prevailing Wind Winter Knots	Summer	Col. 10 Temp. °F Median of Annual Extr. Max.	Min.
TEXAS																
Abilene AP	32 25	99 41	1784	15	20	101/71	99/71	97/71	22	75	74	74	N 12	SSE	103.6	10.4
Alice AP	27 44	98 02	180	31	34	100/78	98/77	95/77	20	82	81	79			104.9	24.8
Amarillo AP	35 14	101 42	3604	6	11	98/67	95/67	93/67	26	71	70	70	N 11	S	100.8	.9
Austin AP	30 18	97 42	597	24	28	100/74	98/74	97/74	22	78	77	77	N 11	S	101.6	19.7
Bay City	29 00	95 58	50	29	33	96/77	94/77	92/77	16	80	79	79				
Beaumont	29 57	94 01	16	27	31	95/79	93/78	91/78	19	81	80	80			99.7	23.5
Beeville	28 22	97 40	190	30	33	99/78	97/77	95/77	18	82	81	79	N 9	SSE	103.1	22.5
Big Spring AP (S)	32 18	101 27	2598	16	20	100/69	97/69	95/69	26	74	73	72			105.3	10.7
Brownsville AP (S)	25 54	97 26	19	35	39	94/77	93/77	92/77	18	80	79	79	NNW 13	SE	98.1	30.1
Brownwood	31 48	98 57	1386	18	22	101/73	99/73	96/73	22	77	76	75	N 9	S	105.3	13.0
Bryan AP	30 40	96 33	276	24	29	98/76	96/76	94/76	20	79	78	78				
Corpus Christi AP	27 46	97 30	41	31	35	95/78	94/78	92/78	19	80	80	79	N 12	SSE	97.0	27.2
Corsicana	32 05	96 28	425	20	25	100/75	98/75	96/75	21	79	78	77			104.2	15.2
Dallas AP	32 51	96 51	481	18	22	102/75	100/75	97/75	20	78	78	77	N 11	S		
Del Rio, Laughlin AFB	29 22	100 47	1081	26	31	100/73	98/73	97/73	24	79	77	76			103.8	23.0
Denton	33 12	97 06	630	17	22	101/74	99/74	97/74	22	78	77	76			104.5	11.8
Eagle Pass	28 52	100 32	884	27	32	101/73	99/73	98/73	24	78	78	77	NNW 9	ESE	107.7	22.1
El Paso AP (S)	31 48	106 24	3918	20	24	100/64	98/64	96/64	27	69	68	68	N 7	S	103.0	15.7
Fort Worth AP (S)	32 50	97 03	537	17	22	101/74	99/74	97/74	22	78	77	76	NW 11	S	103.2	13.5
Galveston AP	29 18	94 48	7	31	36	90/79	89/79	88/78	10	81	80	80	N 15	S	93.9	27.5
Greenville	33 04	96 03	535	17	22	101/74	99/74	97/74	21	78	77	76			103.6	11.7
Harlingen	26 14	97 39	35	35	39	96/77	94/77	93/77	19	80	79	79	NNW 10	SSE	102.3	29.3
Houston AP	29 58	95 21	96	27	32	96/77	94/77	92/77	18	80	79	79	NNW 11	S		
Houston Co	29 59	95 22	108	28	33	97/77	95/77	93/77	18	80	79	79			99.0	23.5
Huntsville	30 43	95 33	494	22	27	100/75	98/75	96/75	20	78	78	77			100.8	18.7
Killeen, Robert Gray AAF	31 05	97 41	850	20	25	99/73	97/73	95/73	22	77	76	75				
Lamesa	32 42	101 56	2965	13	17	99/69	96/69	94/69	26	73	72	71			105.5	8.9
Laredo AFB	27 32	99 27	512	32	36	102/73	101/73	99/74	23	78	78	77	N 8	SE		
Longview	32 28	94 44	330	19	24	99/76	97/76	95/76	20	80	79	78				
Lubbock AP	33 39	101 49	3254	10	15	98/69	96/69	94/69	26	73	72	71	NNE 10	SSE		
Lufkin AP	31 25	94 48	277	25	29	99/76	97/76	94/76	20	80	79	78	NNW 12	S		
Mcallen	26 12	98 13	122	35	39	97/77	95/77	94/77	21	80	79	79				
Midland AP (S)	31 57	102 11	2851	16	21	100/69	98/69	96/69	26	73	72	71	NE 9	SSE	103.6	10.8
Mineral Wells AP	32 47	98 04	930	17	22	101/74	99/74	97/74	22	78	77	76				
Palestine Co	31 47	95 38	600	23	27	100/76	98/76	96/76	20	79	79	78			101.2	16.3
Pampa	35 32	100 59	3250	7	12	99/67	96/67	94/67	26	71	70	70				
Pecos	31 25	103 30	2610	16	21	100/69	98/69	96/69	27	73	72	71				
Plainview	34 11	101 42	3370	8	13	98/68	96/68	94/68	26	72	71	70			102.7	3.1
Port Arthur AP	29 57	94 01	16	27	31	95/79	93/78	91/78	19	81	80	80	N 9	S	97.7	24.0
San Angelo, Goodfellow AFB	31 26	100 24	1877	18	22	101/71	99/71	97/70	24	75	74	73	NNE 10	SSE		
San Antonio AP (S)	29 32	98 28	788	25	30	99/72	97/73	96/73	19	77	76	76	N 8	SSE	101.3	21.1
Sherman, Perrin AFB	33 43	96 40	763	15	20	100/75	98/75	95/74	22	78	77	76	N 10	S	103.0	11.9
Snyder	32 43	100 55	2325	13	18	100/70	98/70	96/70	26	74	73	72				
Temple	31 06	97 21	700	22	27	100/74	99/74	97/74	22	78	77	77				
Tyler AP	32 21	95 16	530	19	24	99/76	97/76	95/76	21	80	79	78	NNE 23	S		
Vernon	34 10	99 18	1212	13	17	102/73	100/73	97/73	24	77	76	75				
Victoria AP	28 51	96 55	104	29	32	98/78	96/77	94/77	18	82	81	79			101.4	23.4
Waco AP	31 37	97 13	500	21	26	101/75	99/75	97/75	22	78	78	77				
Wichita Falls AP	33 58	98 29	994	14	18	103/73	101/73	98/73	24	77	76	75	NNW 12	S		
UTAH																
Cedar City AP	37 42	113 06	5617	-2	5	93/60	91/60	89/59	32	65	63	62	SE 5	SW		
Logan	41 45	111 49	4785	-3	2	93/62	91/61	88/60	33	65	64	63			95.5	-7.8
Moab	38 36	109 36	3965	6	11	100/60	98/60	96/60	30	65	64	63				
Ogden AP	41 12	112 01	4455	1	5	93/63	91/61	88/61	33	66	65	64	S 6	SW	99.5	-3.9
Price	39 37	110 50	5580	-2	5	93/60	91/60	89/59	33	65	63	62				
Provo	40 13	111 43	4448	1	6	98/62	96/62	94/61	32	66	65	64	SE 5	SW		
Richfield	38 46	112 05	5270	-2	5	93/60	91/60	89/59	34	65	63	62			98.1	-10.5
St George Co	37 02	113 31	2900	14	21	103/65	101/65	99/64	33	70	68	67			109.3	11.1
Salt Lake City AP (S)	40 46	111 58	4220	3	8	97/62	95/62	92/61	32	66	65	64	SSE 6	N	99.4	-.1
Vernal AP	40 27	109 31	5280	-5	0	91/61	89/60	86/59	32	64	63	62				
VERMONT																
Barre	44 12	72 31	600	-16	-11	84/71	81/69	78/68	23	73	71	70				
Burlington AP (S)	44 28	73 09	332	-12	-7	88/72	85/70	82/69	23	74	72	71	E 7	SSW	92.4	-16.9
Rutland	43 36	72 58	620	-13	-8	87/72	84/70	81/69	23	74	72	71			92.5	-17.5

Table 3-1. Design temperature criteria, continued. (Copyright 1989 by the American Society of Heating, Refrigerating and Air-Conditioning Engineers, Inc., from the ASHRAE *Handbook—Fundamentals*. Used by permission.)

Col. 1	Col. 2		Col. 3		Col. 4	Col. 5 Winter,[b] °F Design Dry-Bulb		Col. 6 Summer,[c] °F Design Dry-Bulb and Mean Coincident Wet-Bulb			Col. 7 Mean Daily Range	Col. 8 Design Wet-Bulb			Col. 9 Prevailing Wind		Col. 10 Temp. °F Median of Annual Extr.	
State and Station[a]	Lat. ° '		Long. ° '		Elev. Feet	99%	97.5%	1%	2.5%	5%	Range	1%	2.5%	5%	Winter Knots[d]	Summer	Max.	Min.
VIRGINIA																		
Charlottesville	38	02	78	31	870	14	18	94/74	91/74	88/73	23	77	76	75	NE 7	SW	97.4	8.0
Danville AP	36	34	79	20	590	14	16	94/74	92/73	90/73	21	77	76	75			100.1	9.2
Fredericksburg	38	18	77	28	100	10	14	96/76	93/75	90/74	21	78	77	76				
Harrisonburg	38	27	78	54	1370	12	16	93/72	91/72	88/71	23	75	74	73				
Lynchburg AP	37	20	79	12	916	12	16	93/74	90/74	88/73	21	77	76	75	NE 7	SW	97.2	7.6
Norfolk AP	36	54	76	12	22	20	22	93/77	91/76	89/76	18	79	78	77	NW 10	SW	97.2	15.3
Petersburg	37	11	77	31	194	14	17	95/76	92/76	90/75	20	79	78	77				
Richmond AP	37	30	77	20	164	14	17	95/76	92/76	90/75	21	79	78	77	N 6	SW	97.9	9.6
Roanoke AP	37	19	79	58	1193	12	16	93/72	91/72	88/71	23	75	74	73	NW 9	SW		
Staunton	38	16	78	54	1201	12	16	93/72	91/72	88/71	23	75	74	73	NW 9	SW	95.9	2.5
Winchester	39	12	78	10	760	6	10	93/75	90/74	88/74	21	77	76	75			97.3	3.7
WASHINGTON																		
Aberdeen	46	59	123	49	12	25	28	80/65	77/62	73/61	16	65	63	62	ESE 6	NNW	91.9	19.3
Bellingham AP	48	48	122	32	158	10	15	81/67	77/65	74/63	19	68	65	63	NNE 15	WSW	87.4	10.3
Bremerton	47	34	122	40	162	21	25	82/65	78/64	75/62	20	66	64	63	E 8	N		
Ellensburg AP	47	02	120	31	1735	2	6	94/65	91/64	87/62	34	66	65	63				
Everett, Paine AFB	47	55	122	17	596	21	25	80/65	76/64	73/62	20	67	64	63	ESE 6	NNW	84.9	15.2
Kennewick	46	13	119	08	392	5	11	99/68	96/67	92/66	30	70	68	67			103.4	2.0
Longview	46	10	122	56	12	19	24	88/68	85/67	81/65	30	69	67	66	ESE 6	NW	96.0	14.8
Moses Lake, Larson AFB	47	12	119	19	1185	1	7	97/66	94/65	90/63	32	67	66	64	N 8	SSW		
Olympia AP	46	58	122	54	215	16	22	87/66	83/65	79/64	32	67	66	64	NE 4	NE		
Port Angeles	48	07	123	26	99	24	27	72/62	69/61	67/60	18	64	62	61			83.5	19.4
Seattle-Boeing Field	47	32	122	18	23	21	26	84/68	81/66	77/65	24	69	67	65				
Seattle Co (S)	47	39	122	18	20	22	27	85/68	82/66	78/65	19	69	67	65	N 7	N	90.2	22.0
Seattle-Tacoma AP (S)	47	27	122	18	400	21	26	84/65	80/64	76/62	22	66	64	63	E 9	N	90.1	19.9
Spokane AP (S)	47	38	117	31	2357	- 6	2	93/64	90/63	87/62	28	65	64	62	NE 6	SW	98.8	-4.9
Tacoma, McChord AFB	47	15	122	30	100	19	24	86/66	82/65	79/63	22	68	66	64	S 5	NNE	89.4	18.8
Walla Walla AP	46	06	118	17	1206	0	7	97/67	94/66	90/65	27	69	67	66	W 5	W	103.0	3.8
Wenatchee	47	25	120	19	632	7	11	99/67	96/66	92/64	32	68	67	65			101.1	1.0
Yakima AP	46	34	120	32	1052	- 2	5	96/65	93/65	89/63	36	68	66	65	W 5	NW		
WEST VIRGINIA																		
Beckley	37	47	81	07	2504	- 2	4	83/71	81/69	79/69	22	73	71	70	WNW 9	WNW		
Bluefield AP	37	18	81	13	2867	- 2	4	83/71	81/69	79/69	22	73	71	70				
Charleston AP	38	22	81	36	939	7	11	92/74	90/73	87/72	20	76	75	74	SW 8	SW	97.2	2.9
Clarksburg	39	16	80	21	977	6	10	92/74	90/73	87/72	21	76	75	74				
Elkins AP	38	53	79	51	1948	1	6	86/72	84/70	82/70	22	74	72	71	WNW 9	WNW	90.6	-7.3
Huntington Co	38	25	82	30	565	5	10	94/76	91/74	89/73	22	78	77	75	W 6	SW	97.1	2.1
Martinsburg AP	39	24	77	59	556	6	10	93/75	90/74	88/74	21	77	76	75	WNW 10	W	99.0	1.1
Morgantown AP	39	39	79	55	1240	4	8	90/74	87/73	85/73	22	76	75	74				
Parkersburg Co	39	16	81	34	615	7	11	93/75	90/74	88/73	21	77	76	75	WSW 7	WSW	95.9	.7
Wheeling	40	07	80	42	665	1	5	89/72	86/71	84/70	21	74	73	72	WSW 10	WSW	97.5	- .6
WISCONSIN																		
Appleton	44	15	88	23	730	- 14	- 9	89/74	86/72	83/71	23	76	74	72			94.6	- 16.2
Ashland	46	34	90	58	650	- 21	- 16	85/70	82/68	79/66	23	72	70	68			94.1	- 26.8
Beloit	42	30	89	02	780	- 7	- 3	92/75	90/75	88/74	24	78	77	75				
Eau Claire AP	44	52	91	29	888	- 15	- 11	92/75	89/73	86/71	23	77	75	73				
Fond Du Lac	43	48	88	27	760	- 12	- 8	89/74	86/72	84/71	23	76	74	72			96.0	- 17.7
Green Bay AP	44	29	88	08	682	- 13	- 9	88/74	85/72	83/71	23	76	74	72	W 8	SW	94.3	- 17.9
La Crosse AP	43	52	91	15	651	- 13	- 9	91/75	88/73	85/72	22	77	75	74	NW 10	S	95.7	- 21.3
Madison AP (S)	43	08	89	20	858	- 11	- 7	91/74	88/73	85/71	22	77	75	73	NW 8	SW	93.6	- 16.8
Manitowoc	44	06	87	41	660	- 11	- 7	89/74	86/72	83/71	21	76	74	72			94.1	- 13.7
Marinette	45	06	87	38	605	- 15	- 11	87/73	84/71	82/70	20	75	73	71			95.9	- 15.8
Milwaukee AP	42	57	87	54	672	- 8	- 4	90/74	87/73	84/71	21	76	74	73	WNW 10	SSW		
Racine	42	43	87	51	730	- 6	- 2	91/75	88/73	85/72	21	77	75	74				
Sheboygan	43	45	87	43	648	- 10	- 6	89/75	86/73	83/72	20	77	75	74			97.0	- 12.4
Stevens Point	44	30	89	34	1079	- 15	- 11	92/75	89/73	86/71	23	77	75	73			95.3	- 24.1
Waukesha	43	01	88	14	860	- 9	- 5	90/74	87/73	84/71	22	76	74	73			95.7	- 14.3
Wausau AP	44	55	89	37	1196	- 16	- 12	91/74	88/72	85/70	23	76	74	72				
WYOMING																		
Casper AP	42	55	106	28	5338	- 11	- 5	92/58	90/57	87/57	31	63	61	60	NE 10	SW	97.3	- 20.9
Cheyenne	41	09	104	49	6126	- 9	- 1	89/58	86/58	84/57	30	63	62	60	N 11	WNW	92.5	- 15.9
Cody AP	44	33	109	04	4990	- 19	- 13	89/60	86/60	83/59	32	64	63	61			97.4	- 21.9
Evanston	41	16	110	57	6780	- 9	- 3	86/55	84/55	82/54	32	59	58	57			89.2	- 21.2
Lander AP (S)	42	49	108	44	5563	- 16	- 11	91/61	88/61	85/60	32	64	63	61	E 5	NW	94.9	- 22.6
Laramie AP (S)	41	19	105	41	7266	- 14	- 6	84/56	81/56	79/55	28	61	60	59				
Newcastle	43	51	104	13	4265	- 17	- 12	91/64	87/63	84/63	30	69	68	66			99.4	- 19.0
Rawlins	41	48	107	12	6740	- 12	- 4	86/57	83/57	81/56	40	62	61	60				
Rock Springs AP	41	36	109	04	6745	- 9	- 3	86/55	84/55	82/54	32	59	58	57	WSW 10	W		
Sheridan AP	44	46	106	58	3964	- 14	- 8	94/62	91/62	88/61	32	66	65	63	NW 7	N	99.8	- 23.6
Torrington	42	05	104	13	4098	- 14	- 8	94/62	91/62	88/61	30	66	65	63			101.1	- 20.7

Table 3-1. Design temperature criteria, continued. (Copyright 1989 by the American Society of Heating, Refrigerating and Air-Conditioning Engineers, Inc., from the ASHRAE *Handbook—Fundamentals.* Used by permission.)

Solar data

The intensity of solar radiation falling on a surface is a function of the location of the surface, the direction the surface is facing, the clearness of the atmosphere, and the time of year. The intensity of solar radiation reaching a surface outside the earth's atmosphere, perpendicular to the sun's rays, is equal to approximately 430 Btuh/sq ft. The intensity reaching an object on the earth's surface is considerably less.

Solar intensity for a given location and direction is a function of the distance that the surface is from the equator. In the northern hemisphere, the greater the north latitude of the surface, the lower the solar intensity, with locations on the equator being the greatest intensity.

Solar radiation at locations further away from the equator is lower, because the radiation is striking the surface at a sharper angle and must travel a greater distance through the earth's atmosphere.

Solar radiation entering the earth's atmosphere is either reflected (diffused), absorbed, or transmitted through the atmosphere.

Approximately 35% of solar radiation is reflected back into space. Since solar radiation is diffused and absorbed by the earth's atmosphere, radiation falls on surfaces from all parts of the skydome. This partially accounts for the fact that north-facing surfaces, which never receive direct solar radiation, still receive indirect solar radiation.

Ground reflectance also contributes to the intensity of indirect radiation falling on a surface. The total solar intensity falling on surfaces for various north latitudes is given in Table 7-9, Chapter 7.

Solar intensity is also a function of the clearness of the local atmosphere. In industrial and urban areas, solar intensity is reduced due to airborne dust particles, water vapor, carbon dioxide, ozone, and clouds.

Chapter 4

Building Heat Transmission Surfaces

In Chapter 2, the concept of heat transfer for homogeneous objects and surfaces was introduced. This chapter begins the first step in calculating the actual heating and cooling loads for a building. In this chapter, a simplified procedure for estimating the overall conductance through actual typical building surfaces is presented.

The rate of heat transfer per unit area through any given surface is a function of the temperature difference between the two sides and the overall thermal conductance through the composite surface. In order to estimate the heat loss or gain for a building, it is first necessary to determine an overall heat transfer coefficient for each type of surface and the area of each surface exposed to a temperature difference.

Frequently in hvac design, it becomes necessary to calculate the heat transferred through a composite building section rather than through a single element (as has been the case in previous examples). In order to calculate the overall heat transfer through a composite section, the **overall heat transfer coefficient (U value)** has been developed. The U value is a single combined coefficient for conduction through all materials and the convection heat transfer through the surface film.

As an example, suppose a composite brick wall (as shown in Figure 4-1) had an outside temperature of T_o and an inside temperature of T_i.

The U value for the composite wall in Figure 4-1 is a combination of the outside convection coefficient, the thermal conductance of the brick, the thermal conductance of the insulation, the conductance of the drywall, and the inside convection coefficient. The heat transferred through a composite wall can be predicted by applying the following equation:

$$q = (UA) (T_i - T_o)$$

Equation 4.1

where:

q = heat transfer rate (Btuh/sq ft)
U = overall heat transfer coefficient (Btuh/sq ft-°F)
A = wall surface area (sq ft)
T_i = inside air temperature (°F)
T_o = outside air temperature (°F)

The overall heat transfer coefficient is a measure of a composite surface's ability to transmit heat when a temperature difference exists between the two sides.

U value calculation procedure

As the conductance of a material is an indication of the ease with which heat is conducted through the material, the reciprocal of the conductance (the resistance) is a measure of the extent to

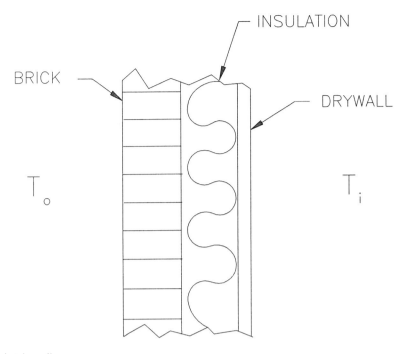

Figure 4-1. Composite brick wall.

which a material resists the flow of heat. Mathematically, the thermal resistance of a material was expressed in Equation 2.4 as:

$$R = \frac{1}{C}$$

The thermal conductance, thermal resistance, and thermal conductivity of many common building materials are presented in Appendix B.

In order to determine the U value of a composite building surface, it is first necessary to determine the thermal resistance of each element, starting with the convection resistance or film resistance for air on either side of the surface. Film resistances for air are given in Table 4-1

After the appropriate film resistances have been selected from Table 4-1, proceed through a cross section of the heat transfer surface, tabulating the thermal resistances for each element of the cross section. The equation for the U value of a composite surface is as follows:

$$U = \frac{1}{R_1 + R_2 + R_3 + \cdots R_n}$$

Equation 4.2

Position of surface	Direction of heat flow	Resistance
Still air		
Horizontal	Upward	0.61
Horizontal	Downward	0.92
Vertical	Horizontal	0.68
Air space	Any	1.01
Moving air		
15-mph wind (winter)	Any	0.17
7.5-mph wind (summer)	Any	0.25

Table 4-1. Air film resistances.

Example 4-1.
Suppose a frame wall had wooden shingles on the outside with felt roll, 1/2-in. plywood, 3-1/2-in. fiberglass batt insulation, and 1/2-in. gypsum board on the inside. Assume there is a 15-mph wind outside. What is the wall's U value?

From Table 4-1 and Appendix B, the following thermal resistances are obtained for each element in the cross section of the wall.

Element	R
Outside air film (15 mph)	0.17
Wooden shingles	0.87
Felt building paper	0.06
1/2-in. plywood	0.63
3-1/2-in. fiberglass insulation	11.83
1/2-in. gypsum board	0.45
Inside air film (horizontal heat flow)	0.68

$$R_t = 14.69$$

$$U = \frac{1}{R_t} = \frac{1}{14.69} = 0.07 \text{ Btuh/sq ft-}°F$$

In Example 4-1, the resultant heat transfer for each square foot of wall area would be 0.07 Btuh for each 1°F temperature difference.

It should be noted that the thermal resistances listed in Appendix B are for a material of a specified thickness. These values may be used directly when the calculation involves materials of the same thickness. Occasionally, materials of different thicknesses from those listed are used. Occasionally, no thickness is tabulated for a material and no corresponding resistance is given. In these cases, it is necessary to use Equation 2.5 to calculate the resistance of the material for the desired thickness. Substituting Equation 2.5 into Equation 2.4 results in Equation 4.3:

$$R = \frac{1}{C} = \left(x \right)\left(\frac{1}{k} \right)$$

Equation 4.3

where:

k = thermal conductivity (Btuh/sq ft-°F-in. or Btuh/sq ft-°F-ft)

x = thickness (in.)

C = conductance of a material

It is important to take care that values with the proper units be used when applying Equation 4.3.

Example 4-2.
Suppose a masonry wall with a 15-mph wind outside is composed of 4-in. face brick, an air space, 8-in. hollow lightweight concrete block,

2-1/2 in. of expanded polystyrene insulation, and 3/4-in. of plaster. What is the wall's U value?

Since the 2-1/2-in. polystyrene insulation does not have a resistance value listed in Appendix B, it is necessary to use Equation 4.3.

For 1-in. polystyrene, Appendix B lists a thermal conductivity of 0.02 Btuh/sq ft-°F.

$$R = \left(x \right)\left(\frac{1}{k} \right) = 2.5 \text{ in.} \left(\frac{1 \text{ ft}}{12 \text{ in.}} \right)\left(\frac{1}{0.02} \right) = 10.42$$

Element	R
Outside air film (15 mph)	0.17
4-in. face brick	0.44
Air space (Table 4-1)	1.01
8-in. hollow lightweight block	2.00
2-1/2-in. polystyrene insulation	10.42
3/4-in. plaster	0.47
Inside air film (horizontal heat flow)	0.68

$$R_t = 15.19$$

$$U = \frac{1}{R_t} = \frac{1}{15.19} = 0.07 \text{ Btuh/sq ft-}°F$$

Example 4-3.
A roof-ceiling consists of 3/8-in. built-up roofing, 3-in. preformed roof insulation, a 20-gauge corrugated metal deck, a 12-in. air space, and 1/2-in. acoustical ceiling tile. What is the U value of the roof-ceiling assembly?

Element	R
Outside air film (15 mph)	0.17
3/8-in. built-up roofing	0.33
3-in. preformed roof insulation	8.33
20-ga corrugated metal deck	0
12-in. air space (Table 4-1)	1.01
1/2-in. acoustical tile	1.26
Inside air film (vertical heat flow)	0.61

$$R_t = 11.71$$

$$U = \frac{1}{R_t} = \frac{1}{11.71} = 0.09 \text{ Btuh/sq ft-}°F$$

In this case, the air space in the ceiling was a "dead" air space; the air was not moving or being exchanged with room air. Frequently in hvac applications, the ceiling space is used as an air plenum to return air from the room back to the air-handling unit. In that case, the space above the ceiling becomes part of the conditioned space, and the resistances for the air space and ceiling tile would not be included in the calculation of the U value for the roof-ceiling assembly.

Note also that the thermal resistance for the metal deck is zero. Metal itself is an excellent conductor of heat and offers little resistance to the flow of heat. For this reason, the thermal resistance of metal can be assumed to be negligible.

In hvac calculations, some types of surfaces, such as glass and various types of doors, are used frequently. Because of this, U values have been tabulated for these surfaces and can be found in Tables 4-2, 4-3, and 4-4.

Thermal mass

In addition to the thermal resistance of a surface, it is also frequently necessary to calculate its thermal mass or weight. Densities for common building materials are presented in Appendix B, along with the materials' thermal conductance or resistance. Since the calculation procedures are similar and the data was obtained from the same tables, it is convenient to calculate the surface weight along with the U value.

In Chapter 2, the section on transient heat transfer discussed the effect of thermal heat storage of materials and non-steady-state heat transfer. The heat storage capacity of a material is closely related to its density or weight. Generally, the greater the density or weight of a material, the greater its capacity to store heat. For this reason, it is usually necessary to determine the weight of an exterior surface.

Description	Exterior Vertical Panels				Exterior Horizontal Panels (Skylights)	
	Summer		Winter		Summer	Winter
	No Indoor Shade	Indoor Shade	No Indoor Shade	Indoor Shade		
Flat Glass						
Single Glass	1.04	0.81	1.10	0.83	0.83	1.23
Insulating Glass, Double						
3/16 in. air space	0.65	0.58	0.62	0.52	0.57	0.70
1/4 in. air space	0.61	0.55	0.58	0.48	0.54	0.65
1/2 in. air space	0.56	0.52	0.49	0.42	0.49	0.59
1/2 in. air space low emittance coating						
$e = 0.20$	0.38	0.37	0.32	0.30	0.36	0.48
$e = 0.40$	0.45	0.44	0.38	0.35	0.42	0.52
$e = 0.60$	0.51	0.48	0.43	0.38	0.46	0.56
Insulating Glass, Triple						
1/4 in. air space	0.44	0.40	0.39	0.31		
1/2 in. air space	0.39	0.36	0.31	0.26		
Storm Windows						
1 in. to 4 in. air spaces	0.50	0.48	0.50	0.42		
Plastic Bubbles						
Single Walled					0.80	1.15
Double Walled					0.46	0.70

Table 4-2. Overall coefficients of heat transmission (U-factor) of windows and skylights Btuh/sq ft-°F. (Copyright 1979 by the American Society of Heating, Refrigerating and Air-Conditioning Engineers, Inc., from the Cooling and Heating Load Calculation Manual. Used by permission.)

Door Thickness, in.	Description	Winter			Summer
		No Storm Door	Wood Storm Door	Metal Storm Door	No Storm Door
1-3/8	Hollow core flush door	0.47	0.30	0.32	0.45
1-3/8	Solid core flush door	0.39	0.26	0.28	0.38
1-3/8	Panel door, 7/16-in. panels	0.57	0.33	0.37	0.54
1-3/4	Hollow core flush door	0.46	0.29	0.32	0.44
	with single glazing	0.56	0.33	0.36	0.54
1-3/4	Solid core flush door	0.33	0.28	0.25	0.32
	with single glazing	0.46	0.29	0.32	0.44
	with insulating glass	0.37	0.25	0.27	0.36
1-3/4	Panel door, 7/16-in. panels	0.54	0.32	0.36	0.52
	with single glazing	0.67	0.36	0.41	0.63
	with insulating glass	0.50	0.31	0.34	0.48
1-3/4	Panel door, 1-1/8-in. panels	0.39	0.26	0.28	0.38
	with single glazing	0.61	0.34	0.38	0.58
	with insulating glass	0.44	0.28	0.31	0.42
2-1/4	Solid core flush door	0.27	0.20	0.21	0.26
	with single glazing	0.41	0.27	0.29	0.40
	with insulating glass	0.33	0.23	0.25	0.32

Table 4-3. Coefficients of transmission (U) for wood doors, Btuh/sq ft-°F. (Copyright 1989 by the American Society of Heating, Refrigerating and Air-Conditioning Engineers, Inc., from the ASHRAE Handbook -- Fundamentals. Used by permission.)

Door Thickness, in.	Description	Winter			Summer
		No Storm Door	Wood Storm Door	Metal Storm Door	No Storm Door
1-3/4	Solid urethane foam core with thermal break	0.19	0.16	0.17	0.18
1-3/4	Solid urethane foam core without thermal break	0.40	-	-	0.39

Table 4-4. Coefficients of transmission (U) for steel doors, Btuh/sq ft-°F. (Copyright 1989 by the American Society of Heating, Refrigerating and Air-Conditioning Engineers, Inc., from the ASHRAE Handbook -- Fundamentals. Used by permission.)

Buildings are frequently classified based on their thermal mass or weight. Thermal mass of a substance or structure is actually the mass times the specific heat. A building with brick and block walls and thick concrete floors is classified as heavy construction. A wooden frame building with wooden floors would be classified as light construction.

Classifications may be based on judgment; however, it is advisable to calculate the weight of surfaces. The mass or weight per unit area of a surface is the best way to classify a surface by weight. The weight per unit area is usually based on 1 sq ft of surface area.

The procedure used to determine the weight of a surface is straightforward. The weight per square foot of surface area of each layer of material is determined by first calculating the volume per square foot for each layer, then multiplying it by the mass density of the material. The weight of each layer is then added together to get the total weight of the surface per square foot.

Buildings are often classified according to the weight of their floors. The classifications are:

- 0 to 64 lb/sq ft, lightweight;
- 65 to 125 lb/sq ft, medium weight;
- Greater than 125 lb/sq ft, heavyweight.

Element	lb/sq ft
Outside air film	0
4-in. face brick; $(4 \text{ in.}) \left(\dfrac{1 \text{ ft}}{12 \text{ in.}} \right) (130 \text{ lb/cu ft})$	43.3
Air space	0
8-in. hollow lightweight block; $(8 \text{ in.}) \left(\dfrac{1 \text{ ft}}{12 \text{ in.}} \right) (45 \text{ lb/cu ft})$	30.0
2-1/2-in. polystyrene insulation; $(2.5 \text{ in.}) \left(\dfrac{1 \text{ ft}}{12 \text{ in.}} \right) (1.8 \text{ lb/cu ft})$	0.4
3/4-in. plaster; $(0.75 \text{ in.}) \left(\dfrac{1 \text{ ft}}{12 \text{ in.}} \right) (50 \text{ lb/cu ft})$	3.1
Inside air film	0
	76.8 lb/sq ft

Example 4-4.
What is the thermal mass or weight per square foot of area of a lightweight (80 lb) concrete floor that is 8-in. thick?

From Appendix B, a density of 80 lb/cu ft is obtained for lightweight concrete that is 8-in. thick. The volume per square foot of surface area is calculated and then multiplied by the density of the material. Therefore, the weight per square foot of floor surface area is determined as follows:

$$\text{Weight} = \left(8 \text{ in.}\right) \left(\frac{1 \text{ ft}}{12 \text{ in.}} \right) \left(80 \text{ lb/cu ft} \right) = 53.33 \text{ lb/sq ft}$$

Example 4-5.
What is the thermal mass or weight of the wall in Example 4-2?

From the tables in Appendix B, we know that:

The weight of the wall is 76.8 lb/sq ft of surface area. The wall is a medium weight surface.

Determining net area of heat transfer

After the different types of heat transfer surfaces have been determined and the respective U value calculated for each, it is necessary to determine the net area through which heat transfer will occur. Heat transfer will occur through any surface that encloses a space that is heated and/or cooled, and the outdoors or an unconditioned space. Examples of an unconditioned space would be an unheated attached garage or a crawl space under the floor of a conditioned space.

The rate of heat transfer through each surface is a function of its U value, the temperature difference across the surface, and the net area of the surface. In order to determine the net area of a heat transfer surface, it is first necessary to determine the gross area of the surface.

The gross area of a wall would consist of all opaque wall areas, window areas, and door areas where the surfaces are exposed to outdoor conditions or an unconditioned space and that enclose a mechanically heated and/or cooled space. This would also include interstitial areas between two conditioned spaces. Similarly, the gross roof-ceiling area would include the entire roof plus any skylights. The net area of each surface is then determined by subtracting the areas of all windows, doors, and other surfaces with different U values.

For the heating-cooling zone of an entire building, the net area is determined for each type of surface for each direction that surface faces. When calculating net heat transfer areas, it is necessary to establish the compass directions on a plan of the space and determine which direction each vertical surface is facing. Net areas must also be calculated for horizontal heat transfer surfaces.

Chapter 5

Infiltration and Ventilation

One of the largest sources of heating and cooling loads for an hvac system is outdoor air. Outdoor air enters and exits a building by one of two means: infiltration-exfiltration and forced ventilation. Infiltration-exfiltration (just infiltration for short) is the random uncontrolled flow or leakage of air into and out of a building. It is important to keep infiltration to a minimum for two reasons:

1. It is undesirable to allow unconditioned air directly into a conditioned space, since this could result in uncomfortable conditions for the occupants.

2. The rate of air exchange is uncontrolled, depending on outdoor conditions such as wind and temperature.

In multi-story buildings, infiltration can also be the result of natural buoyancy forces caused by temperature differences between the inside and outside air (stack effect).

Infiltration can be controlled by the hvac system to some extent, by means of forced ventilation. Infiltration can be reduced by drawing more air into the building through the hvac system than is being mechanically exhausted, thereby pressurizing the building.

The need for outdoor ventilation air in buildings and the effects on indoor air pollution were briefly discussed in Chapter 1. Frequently, the most effective way to control indoor air quality for any given building is by introducing outdoor

air into the building. The amount of outdoor air must be controlled, however. Excess outdoor air ventilation rates can result in higher operating costs, as well as hvac systems with excessive heating and cooling capacity (oversized systems).

Ventilation

The primary need for outdoor ventilation air is to dilute indoor air contaminants. Outdoor air is used to control contaminants such as carbon monoxide, carbon dioxide, odors, formaldehyde, radon, and tobacco smoke. Outdoor ventilation air also may be used to control indoor air humidity.

There is no practical way to remove all of the contaminants from indoor air, especially carbon dioxide. Air filters and cleaners can be used to remove particulate matter from the air but are not effective for gaseous contaminants. Dilution with outdoor air is the most practical method for controlling these indoor air contaminants.

Building mechanical codes set minimum outside air ventilation rates for occupied buildings. The codes govern ventilation of spaces within buildings intended for human occupancy. The most common code has been established by the Building Officials and Code Administration (BOCA).

When discussing ventilation rates, it is necessary to understand how air flows in an hvac system and the terminology used to describe it.

When circulated within a building, air distributed to the various rooms and areas by the air-handling unit is called **supply air ventilation**, or just supply air. Air returned back to the air-handling unit for heating and/or cooling and recirculation is called **return air**. It is normally mixed with air drawn in from outside in the air-handling unit. The supply air is a mixture of conditioned return air and outside air. **Exhaust air** is drawn out of a building mechanically by an exhaust fan. The return air rate is equal to the supply air rate minus the exhaust air rate. The outside air ventilation rate should be equal to or greater than the exhaust air rate.

The BOCA code provides three criteria upon which the ventilation rates for various areas of occupied buildings are determined. The first criteria is a minimum outdoor air ventilation rate per occupant. Presently, the BOCA code requires a minimum outdoor ventilation rate of 15 cfm per person.

A second criteria prescribes the minimum supply air ventilation rate per occupant for various types of occupancies, such as offices, theaters, etc. The code limits the maximum percentage of supply air which may be comprised of recirculated air, depending upon the efficiency of the hvac system's air filtration. The balance of the supply air must be from outside air.

The third criteria of the BOCA code prescribes the minimum amount of air that must be exhausted from spaces such as toilet rooms, locker rooms, janitorial closets, etc. The amount of air that is exhausted must be replaced by the supply air system.

When determining the amount of outside air required to be brought in, or made up, by the hvac system for a building, the designer must examine all three criteria to determine which one governs. The criteria that results in the greatest amount of ventilation is the one that governs.

In addition to consulting local building codes for ventilation requirements, the designer should also consult the latest issue of ASHRAE Standard 62, *Ventilation for Acceptable Indoor Air Quality*. Although ASHRAE Standard 62 is a guide and not a national code at the present time, it is quickly becoming the most accepted guideline for ventilation requirements. Moreover, many local codes now require meeting the criteria of this standard.

In addition to providing guides for minimum outdoor air, the standard provides procedures for ensuring adequate air quality in multiple spaces served by common air systems.

Stack effect

Infiltration and exfiltration are caused by two driving pressures: wind forces and thermal buoyancy (stack effect) due to temperature differences between indoor and outdoor air.

The temperature differences result in pressure differences because of the difference in density between the warm and cold air. Since warm air is less dense than colder air, warm air tends to rise. In winter, warm air tends to rise to the top of the building, creating greater pressure at the top and negative pressure at the bottom. This causes air to exfiltrate through openings and cracks near the top of a tall building and infiltrate at the bottom. A neutral pressure point is reached somewhere near the middle.

Usually, the stack effect is not considered significant for buildings under four stories. For buildings above four stories, the stack effect is only significant in winter, when indoor-outdoor temperature differences are the greatest.

Another factor that tends to mitigate the stack effect is that buildings are not completely open between floors to allow the free flow of air from one level to another. This creates an additional resistance to airflow inside the building, reducing the stack effect. In buildings where the stack effect is considered significant, a rule of thumb for estimating the indoor-outdoor pressure difference due to stack effect is to assume a pressure difference of 0.001 in. of water per story per 1°F temperature difference. (For more detailed calculations of pressure differences due to stack effect, consult References 1 and 4.)

Wind effect

When wind flows over and around an object, such as a building, a relative high pressure is formed outside the windward side of the building and a relative low pressure is formed outside the leeward side. (Pressures on the other sides of the building tend to be neutral, depending on the angle of the wind direction and the shape of the building.)

Pressure differences may vary significantly with time due to air turbulence and changes in wind direction. The relative high pressure on the windward side tends to cause infiltration into the building, while the low pressure causes a corresponding exfiltration on the leeward side.

Combined wind and stack effects

Cracks occur at random fractures in building materials, at the interface of similar or dissimilar materials, and whenever one or more moveable surfaces meet with another surface (windows, doors, etc.). The number and size of the cracks depends on the type of construction, workmanship during installation and manufacture of building components, and on building maintenance after construction.

Air flowing through cracks and other openings in the building envelope is not linearly dependent upon the individual pressure differences caused by the different effects discussed above. Airflows caused by each pressure source should not be added together to get a net infiltration rate.

In order to calculate the infiltration rate, determine the net pressure difference across a surface, then calculate a net infiltration rate. The net pressure difference for any given location in a building may be calculated using the following equation:

$$\Delta P = \Delta P_s + \Delta P_w + \Delta P_p$$

Equation 5.1

where:

ΔP = combined net pressure difference
ΔP_s = pressure difference due to stack effect

ΔP_w = pressure difference due to wind effect
ΔP_p = pressure difference due to pressurization of the building by the hvac system

Each term in Equation 5.1 can be positive or negative, depending on the location within the building and outdoor conditions.

A combined net infiltration rate can then be calculated as follows:

$$Q = (C)(\Delta P)^n$$

Equation 5.2

where:

Q = infiltration-exfiltration airflow rate
ΔP = net pressure difference
n = flow exponent (dependent upon the size and type of crack)
C = pressure coefficient for the crack

The values for calculating pressure differences and infiltration have been determined experimentally for various building components and configurations. Refer to References 1 and 4 for these detailed calculation methods.

Simplified infiltration calculation methods

The most common method for calculating infiltration rates through building openings is the "crack method," which uses tabulated values for infiltration rates in cfm per linear foot of crack length for various types of openings, such as windows and doors. These values assume that the wind is perpendicular to the surface (the wind is blowing directly at the surface). The tables also assume that all cracks around the opening are the same size.

Table 5-1 lists infiltration rates for various types of windows and various qualities of installation or maintenance. Table 5-2 lists infiltration rates for various types of doors and various qualities of installation or maintenance.

| Window type | Infiltration (cfm/ft of crack) | |
	No strip	With strip
Wood sash, double hang		
Average fit	0.65	0.4
Poor fit	1.85	0.57
Poor fit w/storm sash	0.93	0.29
Metal sash, double hang	1.23	0.53
Casement		
Average fit	0.55	—
Poor fit	0.87	—
Vertically pivoted	2.4	—

Table 5-1. Infiltration through windows.

When calculating the rate of infiltration, it is first necessary to determine the crack length for each opening in the surface on the windward side of the building. During winter, this is usually the north side of the building; consult weather data to determine the direction of the prevailing winds.

To determine the required crack length to be used in calculating the rate of infiltration, it is necessary to determine the crack length for both the windward and leeward sides of the building. If air flows into the windward side of the building, an equivalent amount will flow out of the building, mostly from the cracks in the leeward side. If the cracks in the other sides of the building are significantly less than on the windward side, the rate of infiltration could be significantly less, since the building will tend to pressurize. The amount of crack used for calculating infiltration should not be less than one half of the total crack length for the windward side of the room or building.

For small buildings, primarily single or double residences, the air-change method may be used to estimate the rate of infiltration. This simplified rule of thumb method is used for quick estimates and is only applicable to smaller buildings, such as houses. It is based on empirical experience and is applicable to average conditions for average construction.

The method assumes that the air inside the house will be exchanged with air from outside a given number of times per hour on average. The rate at which air infiltrates a building may be determined by multiplying the number of air changes per hour by the total volume of the building. The rate of infiltration can be calculated using Equation 5.3 and Table 5-3.

$$\left(\frac{1 \text{ hr}}{60 \text{ min}}\right)(AC)(V) = (.0167)(AC)(V)$$

Equation 5.3

where:
Q = infiltration rate (cfm)
AC = air changes per hour
V = volume of the house (cu ft)

Door type	Infiltration (cfm/ft of crack)
Wooden or metal	
Good fit w/strip	0.9
Good fit w/o strip	1.8
Poor fit w/o strip	3.7
Glass door	
Good fit	9.6
Average fit	14.0
Poor fit	19.0
Roll-up door	9.6

Table 5-2. Infiltration through doors.

Room type	Air changes per hour	
	Windows with strip	**Windows w/o strip or storm sash**
No exterior windows or doors	0.5	0.3
Windows or doors, 1 side	1	0.7
Windows or doors, 2 sides	1.5	1
Windows or doors, 3 sides or more	2	1.3

Table 5-3. Infiltration air changes.

Heat losses and heat gains due to ventilation and infiltration are discussed further in Chapters 6, 8, and 9.

Chapter 6

Heating Loads

The purpose of a building heating system is to maintain a desired indoor air temperature and indoor comfort conditions despite the loss of heat to the outdoors. The amount of heat loss from a building is directly related to the difference between indoor and outdoor temperatures. A heating load for a building occurs as the result of heat loss or transfer from inside the building to outside.

Heat is lost from a building or space in two ways:

1. Heat is transmitted through the building's exterior surfaces (such as walls, roofs, etc.).

2. Infiltration and/or ventilation exchanges warm indoor air with colder outside air.

The purpose of the heating system is to maintain the desired indoor temperature, say 70°F. In order to do this, the heating system must provide an amount of heat equal to the heat loss from the space at the desired indoor temperature. Figure 6-1 shows a typical heat balance for a space requiring heat.

In order for the temperature of the heated space to remain constant at the desired setpoint, heat supplied by the system must be equal to the combined transmission, ventilation, and infiltration heat losses. The heat loss (combined heat losses) for a space are determined based upon an assumed constant outdoor winter design temperature.

The winter design temperature is based on the climatic conditions for the location of the building and the intended usage of the building. (Climatic conditions and recommended design temperatures were discussed in Chapter 3.)

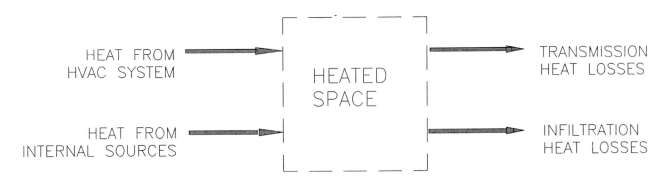

Figure 6-1. Heated space heat balance.

The heating load for a space is usually determined assuming there is no internal heat contribution from internal heat sources, such as lights, people, equipment, etc. The peak heating load usually occurs at night, when outside temperatures are lowest. At night, the building usage is usually quite low and internal heat sources are not occurring. If a credit for some internal heat source is taken, the heat source (heat gain) must be relatively constant. That is, the heat source must occur 24 hours per day.

The credit for the thermal storage of the building itself is also usually not considered, although it can have an effect. In buildings with large thermal mass and daytime internal heat gains, the heating load may be mitigated to some extent.

Transmission heat losses

In order for transmission of heat to occur through a surface, a temperature difference must exist across the surface. As discussed in Chapter 2, heat is transmitted through a substance any time there is a temperature difference between the two sides. The amount of heat transferred is a function of the thermal conductivity of the surface, the temperature difference between the two sides, and the area of the surface.

Transmission of heat from the inside to the outside (a heat loss) occurs through all exterior surfaces of a building and through all surfaces between heated and unheated spaces. An unheated space is any enclosed space adjacent to a heated space that does not have heat directly supplied to it, such as attached unheated garages, attics, crawl spaces, sun porches, etc. The temperature of the unheated space always lies between the temperature of the heated space and the outdoor air temperature.

Example 6-1.
An exterior wall is 10 ft long and 8 ft high with a 4-ft by 4-ft window. The wall has an overall heat transfer coefficient (U value) of 0.05 Btuh/sq ft-°F. The window is a double-pane type with a U value of 0.55 Btuh/sq ft-°F. Assume that the

outside air temperature is -3°F. If the wall and window were the only exterior surfaces in the room and assuming that infiltration is negligible, how much heat would need to be provided to maintain an indoor air temperature of 72°F? (Refer to Chapter 4, Determining net area of heat transfer, where the net surface area for each surface is calculated.) First, determine the window and wall areas in sq ft:

Window area = 4 ft by 4 ft = 16 sq ft

Net wall area = (8 ft by 10 ft) - 16 sq ft (window)
= 64 sq ft

Using Equation 4.1, the heat loss through each surface is calculated as follows:

$$q = (UA)(T_i - T_o)$$

Wall loss: q = (0.05 Btuh/sq ft-°F) (64 sq ft) (72°F - (-3°F)) = 240 Btuh

Glass loss: q = (0.55 Btuh/sq ft-°F) (16 sq ft) (72°F - (-3°F)) = 660 Btuh

Since the infiltration is assumed negligible, the amount of heat necessary to maintain 72°F in the room is:

$$q = 240 \text{ Btuh} + 660 \text{ Btuh} = 900 \text{ Btuh}$$

Transmission heat loss may also occur through surfaces that are indirectly exposed to outdoor temperatures, such as floor slabs and underground walls of basements. In the case of floor slabs and underground walls, the earth acts as an insulator between the inside and outside air. Since the earth can store heat, it acts as a "heat sink" and causes a relatively constant rate of heat loss. (Tables 6-1, 6-2, and 6-3 may be used to calculate heat losses for various below-grade surfaces.) Note that the heat loss for floor slabs is a function of the exposed perimeter of the slab and not the area of the slab.

The heat loss for a floor slab is determined by multiplying the heat loss coefficient for the floor slab, obtained from Table 6-1, by the exposed perimeter of the floor and the indoor-outdoor temperature difference. The degree days (65°F base) needed to use Table 6-1 may be obtained from Reference 5 or a similar source. It may be necessary to interpolate between values listed in Table 6-1 for locations with degree days between the listed values.

Implicit in the above heat loss calculation methods is the assumption of steady-state heat transfer as opposed to non-steady-state, which was discussed in Chapter 2. Because outdoor air temperature fluctuates relatively slowly over time in this case, it is permissible to assume steady-state heat conduction through exterior surfaces when calculating heat losses.

Frequently, daily change in the winter high temperature and low temperature (daily range) results in a temperature change rate of less than 1°F per hour. This is gradual enough to assume steady-state conditions when calculating heat losses.

Infiltration and ventilation heat losses

Whenever colder outside air is introduced into a building, sensible heat from the heating system must be provided to raise the temperature of the outside air to that of the inside air. The addition of latent heat is usually not considered unless the relative humidity in the space is to be maintained at some desired setpoint. Latent heat in the form of moisture may be added with a humidifier.

Applying the First Law of Thermodynamics for a gas, Equation 2.2 gave the amount of heat required to raise the temperature of a substance. In a situation where flow exists, the mass may be given in a mass flow rate such as pounds per

Depth ft	Path Length Through Soil, ft	Heat Loss, Btu/h·ft·°F						
		Uninsulated		R = 4.17		R = 8.34		R = 12.5
0-1	0.68	0.410		0.152		0.093		0.067
1-2	2.27	0.222	0.632	0.116	0.268	0.079	0.172	0.059	0.126
2-3	3.88	0.155	0.767	0.094	0.362	0.068	0.240	0.053	0.179
3-4	5.52	0.119	0.906	0.079	0.441	0.060	0.300	0.048	0.227
4-5	7.05	0.096	1.002	0.069	0.510	0.053	0.353	0.044	0.271
5-6	8.65	0.079	1.081	0.060	0.570	0.048	0.401	0.040	0.311
6-7	10.28	0.069	1.150	0.054	0.624	0.044	0.445	0.037	0.348

Table 6-2. Heat loss below-grade basement walls. (Copyright 1989 by the American Society of Heating, Refrigerating and Air-Conditioning Engineers, Inc., from the ASHRAE Handbook—Fundamentals. Used by permission.)

Depth of Foundation Wall Below Grade	Shortest Width of House, ft			
	20	24	28	32
5 ft	0.032	0.029	0.026	0.023
6 ft	0.030	0.027	0.025	0.022
7 ft	0.029	0.026	0.023	0.021

Note: $\Delta F = (t_a - A)$

Table 6-3. Heat loss through basement floors, Btuh/sq ft-°F. (Copyright 1989 by the American Society of Heating, Refrigerating and Air-Conditioning Engineers, Inc., from the ASHRAE Handbook—Fundamentals. Used by permission.)

Construction	Insulation[a]	Degree Days (65°F Base)		
		2950	5350	7433
8-in. block wall, brick facing	Uninsulated	0.62	0.68	0.72
	R = 5.4 from edge to footer	0.48	0.50	0.56
4-in block wall, brick facing	Uninsulated	0.80	0.84	0.93
	R = 5.4 from edge to footer	0.47	0.49	0.54
Metal stud wall, stucco	Uninsulated	1.15	1.20	1.34
	R = 5.4 from edge to footer	0.51	0.53	0.58
Poured concrete wall with duct near perimeter[b]	Uninsulated	1.84	2.12	2.73
	R = 5.4 from edge to footer, 3 ft under floor	0.64	0.72	0.90

[a]R-value units in °F·ft²·h/Btu·in.

[b]Weighted average temperature of the heating duct was assumed at 110°F during the heating season (outdoor air temperature less than 65°F).

Table 6-1. Heat loss coefficients for floor slab construction, Btuh-°F per ft of perimeter. (Copyright 1989 by the American Society of Heating, Refrigerating and Air-Conditioning Engineers, Inc., from the ASHRAE Handbook—Fundamentals. Used by permission.)

hour or, as is common in hvac design, cubic feet per minute (cfm).

Equation 2.2 can be modified to determine the flow of air required for a space. Air at normal room temperatures has a density of 0.075 lb/cu ft and a specific heat of 0.24 Btu/lb. Equation 2.2 may be rewritten as follows:

$$q = (cfm)(0.075 \text{ lb/cu ft})(0.24 \text{ Btu/lb})(60 \text{ min/hr})(T_2 - T_1)$$

Equation 6.1

where:
q = sensible heat added or removed from the air (Btuh)
cfm = rate of flow (cfm)
$T_2 - T_1$ = temperature change of the air (°F)

Equation 6.1 may be simplified by using a constant for the air density, specific heat, and time conversion and may be written as follows:

$$q = (1.08)(cfm)(T_2 - T_1) \text{ for heating}$$

Equation 6.1a

$$q = (1.10)(cfm)(T_2 - T_1) \text{ for cooling}$$

Equation 6.1b

Methods for determining the rate of infiltration were discussed in Chapter 5. Equation 6.1a may be used to determine heat loss due to infiltration; and Equation 6.1b may be used to determine heat gain.

Example 6-2.
Assume that air infiltrates into a space at the rate of 100 cfm and an equivalent amount of air exfiltrates. The room air temperature is maintained at 70°F; outside air is at 5°F. What is the heat loss due to infiltration? Use Equation 6.1a:

$$q = (1.08)(100 \text{ cfm})(70°F - 5°F) = 7,020 \text{ Btuh}$$

As discussed in Chapter 5, it is frequently mandated by code to introduce outside air into a building via the hvac system. This air also must be heated by the hvac system. It is necessary to determine how much heat must be supplied to a room or space in order to offset heat lost due to infiltration. The infiltration heat loss, along with the transmission heat loss for the space, determines how much air and at what temperature it must be supplied to the space.

In the case of outdoor ventilation air, the heat load is not a part of the individual space or room heat load but part of the hvac system load. Outside air brought into the building by the hvac system is mixed with recirculated or return air, heated or cooled, and supplied to the spaces at the desired temperature.

To determine the sensible heat necessary to raise the total supply air from the temperature of the outside air-return air mixture, it is first necessary to determine the temperature of the mixed air. Equation 6.1a then may be used to determine the heat required to raise the supply air temperature to the desired temperature. The mixed-air temperature may be determined using the following equation:

$$T_m = \frac{(Q_{oa})(T_{oa}) + (Q_{ra})(T_{ra})}{Q_{sa}}$$

Equation 6.2

where:
T_m = temperature of the mixed return air and outside air (°F)
Q_{oa} = volumetric flow rate of the outside air (cfm)
T_{oa} = outside air temperature (°F)
Q_{ra} = volumetric flow rate of the return air (cfm)
T_{ra} = return air temperature (°F); usually same as the room temperature
Q_{sa} = volumetric flow rate of the supply air (cfm); return air + outside air

To determine the heat required by the hvac system, use Equation 6.1a as follows:

$$q = (1.08)(Q_{sa})(T_{sa} - T_m)$$

where:

T_{sa} = required supply air temperature (°F)

The desired supply airflow rate may be determined by code requirements, as discussed in Chapter 5, or by the cooling load, as discussed in Chapter 8. Usually, the required airflow rate to a room is determined by the cooling load. Given a certain airflow rate into a space and the desired room temperature, the required supply air temperature for a given heat load may be determined using Equation 6.3. (Equation 6.3 is obtained by rearranging 6.1a and solving for the supply air temperature.)

$$T_{sa} = T_r + \frac{q}{(1.08)\ (Q_{sa})}$$

Equation 6.3

where:

T_{sa} = required supply air temperature (°F)
T_r = desired room setpoint temperature (°F)
q = total room heat load including transmission and infiltration heat losses (Btuh)
Q_{sa} = volumetric flow rate of supply air (cfm)

Example 6-3.
A room to be maintained at 70°F, with a combined heat loss of 3,000 Btuh, has 100 cfm of air supplied to it. At what temperature must the air be supplied in order to maintain 70°F in the room? Apply Equation 6.3 as follows:

$$T_{sa} = 70°F + \frac{3,000\ Btuh}{(1.08)(100\ cfm)} = 98°F$$

Heating load calculation procedure

A building heating system must be capable of providing sufficient heat in order to maintain the desired inside setpoint temperature. The amount of heat required is equal to the combined heat losses for the space, or entire building, at the desired inside air temperature and the outside design winter conditions.

To determine the required capacity of a heating system, it is necessary to estimate the maximum probable heat loss for each room or space to be heated. The combined heat losses include all transmission heat losses through exterior and underground surfaces and all infiltration heat losses. Ventilation heat loss may be added to the combined space heat losses to get the total combined heat loss. From the total combined heat loss or load, the capacity of the heating system may be determined.

The general procedure for calculating the design heat load for a space or a building is as follows:

1. Select the outdoor design weather conditions based on location and usage of the building.

2. Determine the desired indoor setpoint temperature.

3. Calculate the overall heat transfer coefficient for each heat transfer surface, as described in Chapter 4.

4. Calculate the net heat transfer area of each exterior surface and surfaces adjacent to an unheated space, as described in Chapter 4.

5. Calculate or estimate the rate of infiltration for the space or building, as suggested in Chapter 5.

6. Determine the required outdoor ventilation rate, if any, as described in Chapter 5.

7. Compute the transmission heat loss for each exterior surface and surfaces adjacent to an unconditioned space of the space or building (see Chapter 6).

8. Compute the heat losses from below-grade surfaces such as floor slabs and basement surfaces (refer to Tables 6-1, 6-2, and 6-3).

9. Compute heat losses due to infiltration using Equation 6.1a.

10. Compute the heat losses due to outdoor ventilation (if any) using Equation 6.1a.

11. Compute all transmission, below-grade, infiltration, and ventilation heat losses to get the grand total heat loss for the space or building.

12. Allow for a "pickup load" if night setback will be used. It is estimated that an additional 40% heating capacity should be provided for each 10°F night setback.

13. Select the heating system capacity equal to or slightly greater than the grand total heat loss or pick-up load, whichever is greater.

Example 6-4.
A house in St. Louis, MO has north-south walls that are 30 ft long and east-west walls that are 40 ft long, as shown in Figure 6-2. All walls are 8 ft high. There are two 4 ft by 4 ft double pane windows in the east and west walls. The west wall also has a 3 ft by 7 ft wooden door. The floor slab is a slab-on-grade with no insulation.

Assume that the degree days (65°F base) obtained from tabulated data in Reference 5 is equal to 5,350.

The roof is flat. The west wall is a frame wall, the other three are masonry. The inside temperature is to be maintained at 70°F, except at night when it will be set back to 60°F. Assume that the

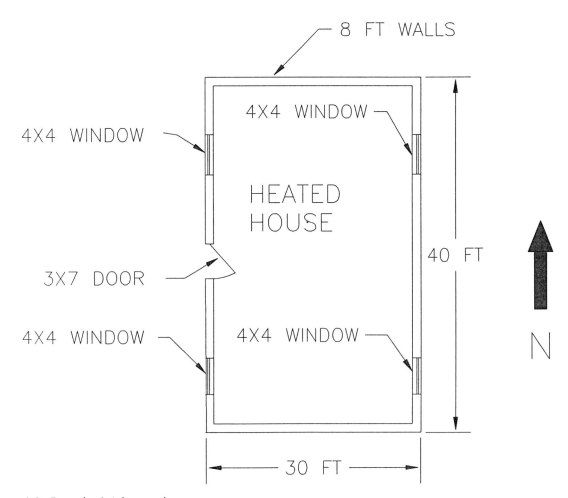

Figure 6-2. Example 6-4 house plan.

infiltration rate of 1/2 air change per hour was calculated by the methods given in Chapter 5.

Assume that the following overall heat transfer coefficients were calculated by the methods given in Chapter 4:

West wall (frame) U = 0.076 Btuh/sq ft-°F

East, south, and
north walls   U = 0.09 Btuh/sq ft-°F

Roof   U = 0.08 Btuh/sq ft-°F

Glass windows   U = 0.56 Btuh/sq ft-°F

Door   U = 0.64 Btuh/sq ft-°F

What is the daytime heat loss for the building, and what size furnace should it have?

Based on the criteria discussed in Chapter 3, an outside temperature is selected from Table 3-1. The outside design temperature selected for St. Louis, MO is 3°F.

The net wall areas are determined by calculating the gross wall area and subtracting the area of all other surfaces in the wall. The net wall areas in square feet are calculated as follows:

	N	E	S	W
Gross wall area				
(length x width)	240	320	240	320
Window area	0	32	0	32
Door area	0	0	0	2
Net wall area	240	288	240	267

Roof area = (40 ft)(30 ft) = 1,200 sq ft
Volume = (1,200 sq ft)(8 ft) = 9,600 cu ft
Floor perimeter (west wall) = 40 ft
Floor perimeter (south, east, and north walls) = 30 ft + 40 ft + 30 ft = 100 ft

Equation 5.3 is used to calculate the rate of infiltration as follows:

$$Q = \frac{(AC/hr)(volume)}{60} = \frac{(0.5)(9,600 \text{ cu ft})}{60 \text{ min/hr}} = 80 \text{ cfm}$$

The transmission losses are calculated using Equation 4.1 as follows:

$$q = (UA)(T_i - T_o)$$

Therefore:

Walls:
West (0.076 Btuh/sq ft-°F)(267 sq ft)(70° - 3°F) = 1,360 Btuh
North (0.09 Btuh/sq ft-°F)(240 sq ft)(70° - 3°F) = 1,447 Btuh
East (0.09 Btuh/sq ft-°F)(288 sq ft)(70° - 3°F) = 1,737 Btuh
South (0.09 Btuh/sq ft-°F)(240 sq ft)(70° - 3°F) = 1,447 Btuh

Windows:
West (0.56 Btuh/sq ft-°F)(32 sq ft)(70° - 3°F) = 1,201 Btuh
East (0.56 Btuh/sq ft-°F)(32 sq ft)(70° - 3°F) = 1,201 Btuh

Door: (0.64 Btuh/sq ft-°F)(21 sq ft)(70° - 3°F) = 900 Btuh

Roof: (0.08 Btuh/sq ft-°F)(1,200 sq ft)(70° - 3°F) = 6,432 Btuh

Referring to Table 6-1, the west frame wall is similar to a metal stud wall and the other three masonry walls are similar to an 8-in. block wall. There is no floor insulation. Heat loss factors for the frame and masonry walls are selected as 1.20 Btuh/ft-°F and 0.68 Btuh/ft-°F, respectively. The transmission heat loss for the floor is as follows:

West floor perimeter:
    (1.20 Btuh/ft-°F)(40 ft)(70° - 3°F) = 3,216 Btuh

Other three walls perimeter:
    (0.68 Btuh/ft-°F)(100 ft)(70° - 3°F) = 4,556 Btuh

Transmission subtotal = 23,497 Btuh

Equation 6.1 and the rate of infiltration calculated above are used to calculate the infiltration heat loss as follows:

Infiltration:
    (1.08)(80 cfm)(70° - 3°F) = 5,789 Btuh

Grand total
heat loss = 23,497 Btuh + 5,789 Btuh = 29,286 Btuh

The grand total heat loss is the heat that would be required to maintain the inside temperature at 70°F with an outside air temperature of 3°F.

Since the temperature is set back, an additional capacity of 40% for each 10°F of set back should be provided for pick-up load:

Allowance
for pick-up = 0.4(29,286 Btuh) = 11,714 Btuh

The required total heating capacity equals 29,286 + 11,714 = 41,000 Btuh.

Therefore, the furnace should have a heating capacity of at least 41,000 Btuh (12 kW) output.

The table on the following page provides step-by-step instructions for performing heat loss calculations.

HEAT LOSS CALCULATIONS

BUILDING: ROOM:

TRANSMISSION LOSS				
ELEMENT	TYPE	U X TEMP. DIFF. (BTUH/SQ FT)	AREA (SQ FT)	LOSS (BTUH)
WALL				
WALL				
WALL				
GLASS				
GLASS				
GLASS				
DOOR				
DOOR				
ROOF/CEILING				
ROOF/CEILING				
SKYLIGHT				
FLOOR				
			SUBTOTAL	

SLAB-ON-GRADE FLOOR			
LENGTH (FT)	FACTOR (BTUH/FT)	LOSS	
TRANSMISSION TOTAL (BTUH)			

INFILTRATION LOSS	
LOSS = 1.08 X °F (TD) X CFM	
SPACE TOTAL LOSS (BTUH)	

Chapter 7

External Heat Gains and Cooling Loads

The purpose of a building's cooling system is to maintain a desired indoor temperature, and sometimes relative humidity level, despite the addition of heat to the space from outdoor conditions and indoor sources. Unlike heating loads, the space cooling load is not just a function of the outdoor air dry bulb temperature. The cooling load for a space or a building is the result of many complex variables, because heat gain to the space occurs from many sources, both inside and outside.

As was discussed in Chapter 2, heat may take one of two forms, sensible heat or latent heat. Sensible heat is associated with a temperature change of a substance, such as air. Latent heat is associated with a phase change of a substance, such as condensation or evaporation. In the case of cooling loads, heat is associated with the evaporation or condensation of moisture or water. For a cooling system, heat is added to the space by evaporation of moisture (such as perspiration) and must be removed as the moisture or latent heat is condensed out of the air.

Basically, heat gain sources may be divided into two major categories: internal heat gain and external heat gain. Internal heat gain results from sources inside the space itself, such as lights, people, equipment, etc. External heat gain comes from sources outside the space or building and result from weather conditions. External heat gains include heat transmission through exterior walls, solar heat gain through windows, and the introduction of outside air into the space in the form of infiltrated warm air. Figure 7-1 depicts a typical heat balance for a space that requires cooling.

Before discussing external heat gains, some definitions of terminology are in order. The distinction should be understood between the instantaneous heat gain for a space, the cooling load for a space, and the heat extraction rate from the space by the cooling system.

The **instantaneous space heat gain** is the rate at which heat is being generated in the space or is entering the space at any given instant in time. In Chapter 2, the effect of building mass and thermal storage on heat gains was discussed.

The **cooling load** is the actual heat transfer to the air in the space at any given time. The space cooling load is different from the space instantaneous heat gain due to the thermal storage characteristics of the space itself. The cooling load is the rate at which heat must be removed from the space to maintain the room dry bulb and wet bulb temperatures at set values.

The **heat extraction rate** is the rate at which heat is actually removed from the space by the hvac system. Due to the nature of hvac systems and temperature control systems, the rate of heat removal is intermittent, causing "swings" in room air temperatures. Most thermostats have a "dead band" temperature range. When the room air temperature is in this range (usually 70° to 78°F), neither heating nor cooling are required. Only

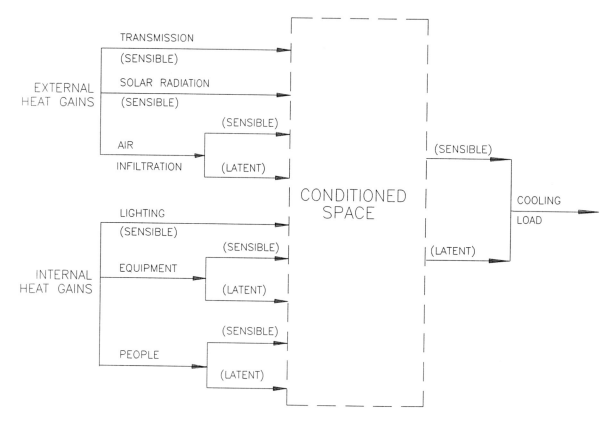

Figure 7-1. A typical heat balance for a space requiring cooling.

when the room air temperature rises above the cooling setpoint temperature (78°F) does the space receive cooling or have heat removed. Heat is removed until the space air temperature drops below the setpoint temperature. This results in a non-constant or varying heat extraction rate, which is frequently different from the actual cooling load at any given time.

External gain calculation methods

The external heat gain or external gain to a space or building is heat gained from outdoors primarily through external surfaces. It includes heat transmission through walls, roofs, doors, windows, etc. and solar heat gain through glass windows and glass doors (fenestration). External heat gain may also result from the addition of heat to a space by the infiltration of outside air into the space.

Unlike heat loss, heat gain through exterior surfaces may not be assumed to be steady state.

Solar radiation on exterior surfaces and the thermal storage affect of the surfaces are the major reasons why external heat gains must be treated as non-steady-state heat transfer. The effect of thermal storage was discussed in Chapter 2.

ASHRAE has developed two calculation methods that take thermal storage and transient heat conduction into account for sunlit exterior surfaces. One method, introduced by ASHRAE in 1972, is the Total Equivalent Temperature Differential (TETD) method. In the TETD method, various components of space heat gain are added together to obtain an instantaneous rate of space heat gain. The space heat gain is converted to an instantaneous cooling load by the Time Averaging (TA) technique, which averages the radiant heat gain with related values for previous hours.

Space heat gains are separated into radiant and convective gains. The convection portion is assumed to be an instantaneous cooling load, while the radiant gain is time averaged. The number of hours the radiant gain is averaged is based upon the thermal mass of the space and exterior surfaces.

The second calculation method is the Transfer Function Method (TFM), which uses a series of "weighing factors" (coefficients of room transfer functions). TFM weighing factors use time averaging to approximate the heat gain for the present hour, and for several previous hours, in order to account for thermal storage of the space and to determine the resultant cooling load.

The actual calculation procedure for the TFM is so complicated that a computer would be necessary to perform the calculations, so a simplified method suitable for manual calculation was developed by ASHRAE. The simplified method uses Cooling Load Temperature Differences (CLTDs) for heat gain through sunlit walls and roofs and for heat conduction through glass (fenestration). The method also uses Cooling Load Factors (CLFs) for solar gains through glass and for internal heat sources. Both CLTDs and CLFs take into account non-steady-state heat transfer caused by thermal storage of the building mass.

CLTDs were calculated and tabulated for various exterior surfaces for three types of room space construction: light, medium, and heavy thermal characteristics. Tables for the various types of wall and roof construction are presented in this chapter.

It is assumed that the heat flow through a similar wall or roof can be determined by multiplying the surface area and overall heat transfer coefficient (U value) by the tabulated CLTD. The tabulated values are based on an inside air temperature of 78°F and air outside design average daily temperature of 85°F. Tabulated values must be adjusted for other indoor air temperatures and other outdoor daily average temperatures. Tabulated values also must be adjusted or corrected for latitude, surface color, and for the month in which the calculations are being performed. A correction factor must also be applied to roofs with various types of attic ventilation.

The methods for manually calculating heat gain and cooling load in this book are based on the

CLTD-CLF method developed by ASHRAE. The procedures presented here are simplified and abridged in order to facilitate understanding the procedures. The reader should understand that calculating cooling loads for a building involves considerable judgment, and the calculations are more of an art than a science.

These procedures will only produce a good estimate of the cooling load for a building. There are a number of limitations involved with the CLTD-CLF method. Most of the limitations are presented here. Please refer to Reference 1 for a complete discussion of the CLTD-CLF method.

Cooling loads for sunlit exterior surfaces

The CLTD method may be used to calculate the cooling load for exterior surfaces, such as walls and roofs, exposed to sunlight and outdoor air temperatures. It is not intended for shaded walls and roofs. The CLTD method accounts for unsteady thermal storage effects caused by varying exterior conditions, such as sunlight and temperature changes. The thermal mass, or weight, of a surface and a procedure to calculate the weight of a surface was discussed in Chapter 4.

Surfaces that have a high mass, such as concrete or brick walls, will have a lower peak space cooling load for a given rate of heat gain. The time at which the peak cooling load occurs also will be delayed, and the cooling load will be spread over a longer period of time.

Surfaces that have the same U value but a light mass, such as a curtain wall or a frame wall, will have a higher peak cooling load, will peak sooner, and will have a shorter period for the cooling load. In an attempt to account for the thermal mass of a surface and transient heat gains, the CLTD is an equivalent steady-state temperature difference that would result in the same cooling load as the surface in question. Representative surface types have been selected and classified or grouped by their weight and U value.

ASHRAE has classified walls by thermal characteristics into seven groups, A through G. Roofs were first divided according to whether the building has or does not have a suspended ceiling. They are then grouped according to their thermal characteristics in groups 1 through 13.

The effect of sunlight on an exterior surface is a function of orientation, or direction the surface is facing, and the time of day. Obviously, the solar heat gain for an east-facing wall at 9:00 a.m. will be much greater than at 6:00 p.m. In addition to the time of day, the peak heat gain is a function of the day of the month, the season of the year, and the location of the building with respect to the earth's equator (latitude).

Finally, the peak heat gain is also a function of the orientation of the surface (horizontal, vertical, sloping, east facing, south facing, etc.). The amount and intensity of solar radiation is a function of all of the variables given above.

In order to account for the effect of varying solar radiation on a surface and the mass of a surface, the "Sol-air" temperature concept was developed. The Sol-air temperature is that temperature of the outdoor air which, in the absence of all radiation exchanges, would result in the same rate of heat gain to the surface as would exist with the actual combination of incident solar radiation, radiant energy exchange with the sky, and other outdoor surroundings and convective air.

The CLTD is similar to the Sol-air temperature in that it attempts to predict a temperature difference and cooling load for exterior surfaces that would result in the presence of solar radiation. It also attempts to account for the thermal mass of a surface and variation of the cooling load with time.

In its simplified form, Equation 7.1 may be used to predict the cooling load for a given surface:

$$q = (UA)(CLTD)$$

Equation 7.1

where:

q = cooling load (Btuh)
U = overall heat transfer coefficient (Btuh/sq ft-°F)
A = area of surface (sq ft)
CLTD = temperature difference which gives the cooling load at the designated time for the given surface type

The CLTD for actual situations must be corrected for individual situations, as will be discussed in the following sections.

To properly use the CLTD method, the time of day at which the peak load occurs must be estimated. As will be discussed in the next section, this requires considerable judgment since different surfaces in a conditioned space frequently peak at different times. An east wall, a west wall, and a roof each have a different time at which the peak CLTD occurs. In addition, rooms with different orientations in the building may each peak at a different time. The actual peak load for a cooling system serving several rooms may occur at a different time than any individual room.

Cooling loads for sunlit walls

ASHRAE has classified or grouped exterior walls according to their construction weight and U value. There are seven different wall groups, A through G. Table 7-1 lists the seven wall types, along with typical representative wall construction types and typical U values for each group. It can be observed that Group A walls have a relatively high mass or weight, while Group G walls are lightweight.

The first step in the cooling load calculation procedure is to select a wall group from Table 7-1 that is similar in weight to the wall for which load calculations are being made. After a similar wall construction group is selected, the appropriate CLTD may be obtained from Table 7-2, where seven different wall group CLTDs are listed depending upon the direction the wall is facing (orientation) and the time of day. The times that are listed are solar times. Notice that for any given orientation, the heavier Group A

Group No.	Description of Construction	Weight (lb/ft²)	U-Value (Btu/h·ft²·°F)
4-in. Face brick + (brick)			
C	Air space + 4-in. face brick	83	0.358
D	4-in. common brick	90	0.415
C	1-in. insulation or air space + 4-in. common brick	90	0.174-0.301
B	2-in. insulation + 4-in. common brick	88	0.111
B	8-in. common brick	130	0.302
A	Insulation or air space + 8-in. common brick	130	0.154-0.243
4-in. Face brick + (heavyweight concrete)			
C	Air space + 2-in. concrete	94	0.350
B	2-in. insulation + 4-in. concrete	97	0.116
A	Air space or insulation + 8-in. or more concrete	143-190	0.110-0.112
4-in. Face brick + (light or heavyweight concrete block)			
E	4-in. block	62	0.319
D	Air space or insulation + 4-in. block	62	0.153-0.246
D	8-in. block	70	0.274
C	Air space or 1-in. insulation + 6-in. or 8-in. block	73-89	0.221-0.275
B	2-in. insulation + 8-in. block	89	0.096-0.107
4-in. Face brick + (clay tile)			
D	4-in. tile	71	0.381
D	Air space + 4-in. tile	71	0.281
C	Insulation + 4-in. tile	71	0.169
C	8-in. tile	96	0.275
B	Air space or 1-in. insulation + 8-in. tile	96	0.142-0.221
A	2-in. insulation + 8-in. tile	97	0.097
Heavyweight concrete wall + (finish)			
E	4-in. concrete	63	0.585
D	4-in. concrete + 1-in. or 2-in. insulation	63	0.119-0.200
C	2-in. insulation + 4-in. concrete	63	0.119
C	8-in. concrete	109	0.490
B	8-in. concrete + 1-in. or 2-in. insulation	110	0.115-0.187
A	2-in. insulation + 8-in. concrete	110	0.115
B	12-in. concrete	156	0.421
A	12-in. concrete + insulation	156	0.113
Light and heavyweight concrete block + (finish)			
F	4-in. block + air space/insulation	29	0.161-0.263
E	2-in. insulation + 4-in. block	29-37	0.105-0.114
E	8-in. block	47-51	0.294-0.402
D	8-in. block + air space/insulation	41-57	0.149-0.173
Clay tile + (finish)			
F	4-in. tile	39	0.419
F	4-in. tile + air space	39	0.303
E	4-in. tile + 1-in. insulation	39	0.175
D	2-in. insulation + 4-in. tile	40	0.110
D	8-in. tile	63	0.296
C	8-in. tile + air space/1-in. insulation	63	0.151-0.231
B	2-in. insulation + 8-in. tile	63	0.099
Metal curtain wall			
G	With/without air space + 1- to 3-in. insulation	5-6	0.091-0.230
Frame wall			
G	1-in. to 3-in. insulation	16	0.081-0.178

Table 7-1. Wall construction groups. (Copyright 1989 by the American Society of Heating, Refrigerating and Air-Conditioning Engineers, Inc., from the ASHRAE *Handbook—Fundamentals*. Used by permission.)

North Latitude Wall Facing	0100	0200	0300	0400	0500	0600	0700	0800	0900	1000	1100	1200	1300	1400	1500	1600	1700	1800	1900	2000	2100	2200	2300	2400	Hr of Maximum CLTD	Minimum CLTD	Maximum CLTD	Difference CLTD
Group A Walls																												
N	14	14	14	13	13	13	12	12	11	11	10	10	10	10	10	10	11	11	12	12	13	13	14	14	2	10	14	4
NE	19	19	19	18	17	17	16	15	15	15	15	15	16	16	17	18	18	18	19	19	20	20	20	20	22	15	20	5
E	24	24	23	23	22	21	20	19	19	18	19	19	20	21	22	23	24	24	25	25	25	25	25	22	18	18	25	7
SE	24	23	23	22	21	20	20	19	18	18	18	18	18	19	20	21	22	23	23	24	24	24	24	22	18	18	24	6
S	20	20	19	19	18	18	17	16	16	15	14	14	14	14	14	15	16	17	18	19	19	20	20	20	23	14	20	6
SW	25	25	25	24	24	23	22	21	20	19	19	18	17	17	17	17	18	19	20	22	23	24	25	25	24	17	25	8
W	27	27	26	26	25	24	24	23	22	21	20	19	19	18	18	18	18	19	20	22	23	25	26	26	1	18	27	9
NW	21	21	21	20	20	19	19	18	17	16	16	15	15	14	14	14	15	15	16	17	18	19	20	21	1	14	21	7
Group B Walls																												
N	15	14	14	13	12	11	11	10	9	9	9	8	9	9	9	10	11	12	13	14	14	15	15	15	24	8	15	7
NE	19	18	17	16	15	14	13	12	12	13	14	15	16	17	18	19	19	20	20	21	21	21	20	20	21	12	21	9
E	23	22	21	19	18	17	16	15	15	15	17	19	21	22	24	25	26	26	27	27	26	26	25	20	15	15	27	12
SE	23	22	21	20	18	17	16	15	14	14	15	16	18	20	21	23	24	25	26	26	26	25	24	21	14	14	26	12
S	21	20	19	18	17	15	14	13	12	11	11	11	12	14	15	17	19	20	21	22	22	22	21	23	11	11	22	11
SW	27	26	25	24	22	21	19	18	16	15	14	14	13	13	14	15	17	20	22	25	27	28	28	24	13	13	28	15
W	29	28	27	26	24	23	21	19	18	17	16	15	14	14	14	15	17	19	22	25	27	29	29	30	24	14	30	16
NW	23	22	21	20	19	18	17	15	14	13	12	12	12	11	12	12	13	15	17	19	21	22	23	23	24	11	23	9
Group C Walls																												
N	15	14	13	12	11	10	9	8	8	7	7	8	8	9	10	12	13	14	15	16	17	17	17	16	22	7	17	10
NE	19	17	16	14	13	11	10	10	11	13	15	17	19	20	21	22	23	23	23	23	22	21	20	20	20	10	23	13
E	22	21	19	17	15	14	12	12	14	16	19	22	25	27	29	30	30	30	29	28	27	26	24	18	12	12	30	18
SE	22	21	19	17	15	14	12	12	12	13	16	19	22	24	26	28	29	29	29	29	28	27	26	24	19	12	29	17
S	21	19	18	16	15	13	12	10	9	9	9	10	11	14	17	20	22	24	25	25	25	24	22	20	9	9	26	17
SW	29	27	25	22	20	18	16	15	13	12	11	11	11	13	15	18	22	26	29	32	33	33	32	31	22	11	33	22
W	31	29	27	25	22	20	18	16	14	13	12	12	12	13	14	16	20	24	29	32	35	35	35	33	22	12	35	23
NW	25	23	21	20	18	16	14	13	11	10	10	10	10	11	12	13	15	18	22	25	27	27	27	26	22	10	27	17
Group D Walls																												
N	15	13	12	10	9	7	6	6	6	6	6	7	8	10	12	13	15	17	18	19	19	19	18	16	21	6	19	13
NE	17	15	13	11	10	8	7	8	10	14	17	20	22	23	23	24	24	25	25	24	23	22	20	18	19	7	25	18
E	19	17	15	13	11	9	8	9	12	17	22	27	30	32	33	33	32	32	31	30	28	26	24	22	16	8	33	25
SE	20	17	15	13	11	10	8	8	10	13	17	22	26	29	31	32	32	32	31	30	28	26	24	22	17	8	32	24
S	19	17	15	13	11	9	8	7	6	6	7	9	12	16	20	24	27	29	29	29	27	26	24	22	19	6	29	23
SW	28	25	22	19	16	14	12	10	9	8	8	8	10	12	16	21	27	32	36	38	38	37	34	31	21	8	38	30
W	31	27	24	21	18	15	13	11	10	9	9	9	10	11	14	18	24	30	36	40	41	40	38	34	21	9	41	32
NW	25	22	19	17	14	12	10	9	8	7	7	8	9	10	12	14	18	22	27	31	32	32	30	27	22	7	32	25
Group E Walls																												
N	12	10	8	7	5	4	3	4	5	6	7	9	11	13	15	17	19	20	21	23	20	18	16	14	20	3	22	19
NE	13	11	9	7	6	4	5	9	15	20	24	25	25	26	26	26	26	24	22	19	17	15	16	14	16	4	26	22
E	14	12	10	8	6	5	6	11	18	26	33	36	38	37	36	34	33	32	30	28	25	22	20	17	13	5	38	33
SE	15	12	10	8	7	5	5	8	12	19	25	31	35	37	37	36	34	33	31	28	26	23	20	17	15	5	37	32
S	15	12	10	8	7	5	4	4	5	9	13	19	24	29	32	34	33	31	29	26	23	20	17	17	15	3	34	31
SW	22	18	15	12	10	8	6	5	5	6	7	9	12	18	24	32	38	43	45	44	40	35	30	26	19	5	45	40
W	25	21	17	14	11	9	7	6	6	6	7	9	11	14	20	27	36	43	49	49	45	40	34	29	20	6	49	43
NW	20	17	14	11	9	7	6	5	5	5	6	8	10	13	16	20	26	32	37	38	36	32	28	24	20	5	38	33
Group F Walls																												
N	8	6	5	3	2	1	2	4	6	7	9	11	14	17	19	21	22	23	24	23	20	16	13	11	19	1	23	23
NE	9	7	5	3	2	1	5	14	23	28	30	29	28	27	27	26	24	22	19	16	13	11	11	11	11	1	30	29
E	10	7	6	4	3	2	6	17	28	38	44	45	43	39	36	34	32	30	27	24	21	17	15	12	12	2	45	43
SE	10	7	6	4	3	2	4	10	19	28	36	41	43	42	39	36	34	31	28	25	21	18	15	12	13	2	43	41
S	10	8	6	4	3	2	1	1	3	7	13	20	27	34	38	39	38	35	31	26	22	18	15	12	16	1	39	38
SW	15	11	9	6	5	3	2	2	4	5	8	11	17	26	35	44	50	53	52	45	37	28	23	18	18	2	53	48
W	17	13	10	7	5	4	3	3	4	6	8	11	14	20	28	39	49	57	60	54	43	34	27	21	19	3	60	57
NW	14	10	8	6	4	3	2	2	3	5	8	10	13	15	21	27	35	42	46	43	35	28	22	18	19	2	46	44
Group G Walls																												
N	3	2	1	0	-1	2	7	8	9	12	15	18	21	23	24	24	25	26	22	15	11	9	7	5	18	-1	26	27
NE	3	2	1	0	-1	9	27	36	39	35	30	26	26	27	27	26	25	22	18	14	11	9	7	5	9	-1	39	40
E	4	2	1	0	-1	11	31	47	54	55	50	40	33	31	30	29	27	24	19	15	12	10	8	6	10	-1	55	56
SE	4	2	1	0	-1	5	18	32	42	49	51	48	42	36	30	27	24	19	15	12	10	8	6	6	11	-1	51	52
S	4	2	1	0	-1	0	1	5	12	22	31	39	45	46	43	37	31	25	20	15	12	10	8	5	14	-1	46	47
SW	5	4	3	1	0	0	2	5	8	12	16	26	38	50	59	63	61	52	37	24	17	13	10	8	16	0	63	63
W	6	5	3	2	1	1	2	5	8	11	15	19	27	41	56	67	72	67	48	29	20	15	11	8	17	1	72	71
NW	5	3	2	1	0	0	2	5	8	11	15	18	21	27	37	47	55	55	41	25	17	13	10	7	18	1	55	55

(1) *Direct Application of the Table Without Adjustments:*

Values in this table were calculated using the same conditions for walls as outlined for the roof CLTD table, These values may be used for all normal air-conditioning estimates usually without correction (except as noted below) when the load is calculated for the hottest weather.

For totally shaded walls use the North orientation values.

(2) *Adjustments to Table Values:*

The following equation makes adjustments for conditions other than those listed in Note (1).

$$CLTD_{corr} = (CLTD + LM)\, K + (78 - t_R) + (t_o - 85)$$

where
CLTD is from Table 31 at the wall orientation.
LM is the latitude-month correction from Table 32.

K is a color adjustment factor applied after first making month-latitude adjustment
$K = 1.0$ if dark colored or light in an industrial area
$K = 0.83$ if permanently medium-colored (rural area)
$K = 0.65$ if permanently light-colored (rural area)
Credit should not be taken for wall color other than dark except where permanence of color is established by experience, as in rural areas or where there is little smoke.

Colors:
Light — Cream
Medium — Medium blue, medium green, bright red, light brown, unpainted wood and natural color concrete
Dark — Dark blue, red, brown and green

$(78 - t_R)$ is indoor design temperature correction
$(t_o - 85)$ is outdoor design temperature correction, where t_o is the average outside temperature on design day.

Table 7-2. Cooling load temperature differences for sunlit walls. (Copyright 1989 by the American Society of Heating, Refrigerating and Air-Conditioning Engineers, Inc., from the ASHRAE *Handbook—Fundamentals*. Used by permission.)

walls have a lower peak CLTD, and that peak occurs at a later hour than the lightweight Group G walls.

The cooling load temperature differences listed in Table 7-2 are based upon several assumed conditions:

- The solar radiation is for 40-degrees north latitude on July 21.
- The surface is assumed to be dark in color.
- The inside air temperature is 78°F.
- The outside air temperature is 95°F with an outdoor mean temperature of 85°F and a daily range (variance) of 21°F.
- The outside air film resistance is 0.333 hr-sq ft-°F/Btu.
- The inside air film resistance is 0.685 hr-sq ft-°F/Btu.

If the design conditions for the building under consideration differ from the above conditions, the CLTD values must be modified or corrected before using Equation 7.1 to calculate the cooling load for the surface.

The following equation may be used to correct the CLTD values obtained from Table 7-2:

$$CLTD_c = (CLTD + LM)(k) + (78°F - T_i) + (T_{oa} - 85°F)$$

Equation 7.2

where:
$CLTD_c$ = corrected cooling load temperature difference
$CLTD$ = cooling load temperature difference obtained from Table 7-2
LM = latitude and month correction for the given surface location and calculation month. (LM is obtained from Table 7-3, CLTD correction for latitude and month.)
k = color adjustment factor ($k = 1$ if the surface is a dark color; $k = 0.5$ if the surface is permanently light colored and is in a rural area where there is little smoke)
T_i = indoor design air temperature (°F)

T_{oa} = average outside air temperature on the design day (T_{oa} = outside design temperature minus one half the daily range)

Example 7-1.

Assume there is a 50-sq ft west-facing wall, located at 24-degrees north latitude on July 21. The wall has a 4-in. medium-weight hollow concrete block exterior, a 1-in. air space, 1-in. expanded polystyrene insulation, and 3/4-in. gypsum board interior finish. The outside design dry bulb temperature is 95°F, the daily range is 20°F, and the inside design dry bulb temperature is 75°F. What is the peak cooling load for the wall, and at what hour does it occur?

First it is necessary to calculate the U value and weight of the wall (refer to Chapter 4 and Appendix B).

Element	R	Weight/ sq ft
Outside air film (7.5 mph)	0.25	0
4-in. medium-weight hollow concrete block	1.11	25.33
1-in. air space	0.89	0
1-in. expanded polystyrene	4.16	0.15
3/4-in. gypsum board	0.67	3.13
Inside air film	0.68	0
	7.76	28.61

$$U = \frac{1}{7.76} = 0.13 \text{ Btuh/sq ft-°F}$$

Weight = 28.61 lb/sq ft

Referring to Table 7-1, a Type F wall with 4-in. lightweight or heavyweight concrete block + air space-insulation + finish is similar. The representative wall has a weight of 29 lb/sq ft and a U value between 0.161 and 0.263.

Therefore, the calculations will be based on a Type F wall with a U value of 0.13 Btuh/sq ft-°F. Assume the wall is dark, since it is not permanently light colored; therefore, $k = 1$.

From Table 7-2 it is determined that a Group F west-facing wall has a maximum or peak CLTD of 60°F, and it occurs at hour 19:00. The CLTD must now be corrected for the latitude-month, outside design temperature, and inside design temperature. From Table 7-3, the correction for a west-facing wall at 24-degrees north latitude on July 21 is 0.

For an outside design dry bulb temperature of 95°F and a daily range of 20°F, the average daily dry bulb temperature is:

$$T_{oa} = 95°F - \frac{20°F}{2} = 85°F$$

Using Equation 7.2 for an inside design dry bulb temperature of 75°F, the corrected CLTD is:

$$CLTD_c = (60°F + 0)(1) + (78°F - 75°F) + (85°F - 85°F) = 63°F$$

Using Equation 7.1, the peak cooling load is:

$$q = (0.13 \text{ Btuh/sq ft-°F})(50 \text{ sq ft})(63°F) = 409 \text{ Btuh}$$

When calculating cooling loads for spaces, considerable judgment is required. First, an exterior surface must be selected from the tables that has a similar weight and U value as the design surface. In all cases, the actual U value of the design surface should be used in Equation 7.1 to calculate the cooling load for the surface.

Secondly, if there is more than one exterior surface, or heat source for that matter, considerable judgment must be used to select the hour at which the peak load for the space or room occurs. For this reason, it is frequently necessary to calculate cooling loads for several different hours in order to determine the actual peak cooling load for a space and the time at which it occurs.

Example 7-2.

A room has two 50-sq ft walls that have the same construction as the wall described in Example 7-1. One wall faces east and one faces south. Assume the same location and design temperatures as Example 7-1. What is the peak cooling load for the space and when does it occur?

From Table 7-2, CLTD values and times are obtained as follows:

	12:00	14:00	16:00
East wall	45	39	34
South wall	20	34	39

It can be seen that the east wall has a peak CLTD of 45°F at hour 12:00, and then the CLTD decreases. The south wall CLTD continues to increase after hour 12:00 until it peaks at a value of 39°F at hour 16:00. Since the two surfaces peak at different times, it will be necessary to calculate the cooling load for several different hours.

Using Equation 7.2, the corrected CLTD values for the east wall are as follows:

At hour 12:00 —

$$CLTD_c = (45°F + 0)(1) + (78°F - 75°F) + (85°F - 85°F) = 48°F$$

At hour 14:00 —

$$CLTD_c = (39°F + 0)(1) + (78°F - 75°F) + (85°F - 85°F) = 42°F$$

At hour 16:00 —

$$CLTD_c = (34°F + 0)(1) + (78°F - 75°F) + (85°F - 85°F) = 37°F$$

For the west wall, the corrected CLTD values are as follows:

At hour 12:00 —

$$CLTD_c = (20°F + 0)(1) + (78°F - 75°F) + (85°F - 85°F) = 23°F$$

At hour 14:00 —

$$CLTD_c = (34°F + 0)(1) + (78°F - 75°F) + (85°F - 85°F) = 37°F$$

Lat.	Month	N	NNE NNW	NE NW	ENE WNW	E W	ESE WSW	SE SW	SSE SSW	S	HOR
0	Dec	-3	-5	-5	-5	-2	0	3	6	9	-1
	Jan/Nov	-3	-5	-4	-4	-1	0	2	4	7	-1
	Feb/Oct	-3	-2	-2	-2	-1	-1	0	-1	0	0
	Mar/Sept	-3	0	1	-1	-1	-3	-3	-5	-8	0
	Apr/Aug	5	4	3	0	-2	-5	-6	-8	-8	-2
	May/Jul	10	7	5	0	-3	-7	-8	-9	-8	-4
	Jun	12	9	5	0	-3	-7	-9	-10	-8	-5
8	Dec	-4	-6	-6	-6	-3	0	4	8	12	-5
	Jan/Nov	-3	-5	-6	-5	-2	0	3	6	10	-4
	Feb/Oct	-3	-4	-3	-3	-1	-1	1	2	4	-1
	Mar/Sept	-3	-2	-1	-1	-1	-2	-2	-3	-4	0
	Apr/Aug	2	2	2	0	-1	-4	-5	-7	-7	-1
	May/Jul	7	5	4	0	-2	-5	-7	-9	-7	-2
	Jun	9	6	4	0	-2	-6	-8	-9	-7	-2
16	Dec	-4	-6	-8	-8	-4	-1	4	9	13	-9
	Jan/Nov	-4	-6	-7	-7	-4	-1	4	8	12	-7
	Feb/Oct	-3	-5	-5	-4	-2	0	2	5	7	-4
	Mar/Sept	-3	-3	-2	-2	-1	-1	0	0	0	-1
	Apr/Aug	-1	0	-1	-1	-1	-3	-3	-5	-6	0
	May/Jul	4	3	3	0	-1	-4	-5	-7	-7	0
	Jun	6	4	4	1	-1	-4	-6	-8	0	-7
24	Dec	-5	-7	-9	-10	-7	-3	3	9	13	-13
	Jan/Nov	-4	-6	-8	-9	-6	-3	9	3	13	-11
	Feb/Oct	-4	-5	-6	-6	-3	-1	3	7	10	-7
	Mar/Sept	-3	-4	-3	-3	-1	-1	1	2	4	-3
	Apr/Aug	-2	-1	0	-1	-1	-2	-1	-2	-3	0
	May/Jul	1	2	2	0	0	-3	-3	-5	-6	1
	Jun	3	3	3	1	0	-3	-4	-6	-6	1
32	Dec	-5	-7	-10	-11	-8	-5	2	9	12	-17
	Jan/Nov	-5	-7	-9	-11	-8	-15	-4	2	9	12
	Feb/Oct	-4	-6	-7	-8	-4	-2	4	8	11	-10
	Mar/Sept	-3	-4	-4	-4	-2	-1	3	5	7	-5
	Apr/Aug	-2	-2	-1	-2	0	-1	0	1	1	-1
	May/Jul	1	1	1	0	0	-1	-1	-3	-3	1
	Jun	1	2	2	1	0	-2	-2	-4	-4	2
40	Dec	-6	-8	-10	-13	-10	-7	0	7	10	-21
	Jan/Nov	-5	-7	-10	-12	-9	-6	1	8	11	-19
	Feb/Oct	-5	-7	-8	-9	-6	-3	3	8	12	-14
	Mar/Sept	-4	-5	-5	-6	-3	-1	4	7	10	-8
	Apr/Aug	-2	-3	-2	-2	0	0	2	3	4	-3
	May/Jul	0	0	0	0	0	0	0	0	1	1
	Jun	1	1	1	0	1	0	0	-1	-1	2
48	Dec	-6	-8	-11	-14	-13	-10	-3	2	6	-25
	Jan/Nov	-6	-8	-11	-13	-11	-8	-1	5	8	-24
	Feb/Oct	-5	-7	-10	-11	-8	-5	1	8	11	-18
	Mar/Sept	-4	-6	-6	-7	-4	-1	4	8	11	-11
	Apr/Aug	-3	-3	-3	-3	-1	0	4	6	7	-5
	May/Jul	0	-1	0	0	1	1	3	3	4	0
	Jun	1	1	2	1	2	1	2	2	3	2
56	Dec	-7	-9	-12	-16	-16	-14	-9	-5	-3	-28
	Jan/Nov	-6	-8	-11	-15	-14	-12	-6	-1	2	-27
	Feb/Oct	-6	-8	-10	-12	-10	-7	0	6	9	-22
	Mar/Sept	-5	-6	-7	-8	-5	-2	4	8	12	-15
	Apr/Aug	-3	-4	-4	-4	-1	1	5	7	9	-8
	May/Jul	0	0	0	0	2	2	5	6	7	-2
	Jun	2	1	2	1	3	3	4	5	6	1
64	Dec	-7	-9	-12	-16	-17	-18	-16	-14	-12	-30
	Jan/Nov	-7	-9	-12	-16	-16	-16	-13	-10	-8	-29
	Feb/Oct	-6	-8	-11	-14	-13	10	-4	1	4	-26·
	Mar/Sept	-5	-7	-9	-10	-7	-4	2	7	11	-20
	Apr/Aug	-3	-4	-4	-4	-1	1	5	9	11	-11
	May/Jul	1	0	1	0	3	4	6	8	10	-3
	Jun	2	2	2	2	4	4	6	7	9	0

Table 7-3. CLTD correction for latitude and month. (Copyright 1989 by the American Society of Heating, Refrigerating and Air-Conditioning Engineers, Inc., from the ASHRAE *Handbook—Fundamentals*. Used by permission.)

At hour 16:00 —

$$CLTD_c = (39°F + 0)(1) + (78°F - 75°F) + (85°F - 85°F)$$
$$= 42°F$$

The total external cooling load for the space is the sum of the cooling loads for each exterior surface. The cooling load at each hour is as follows:

At hour 12:00 —

$$q = (0.13 \text{ Btuh/sq ft-°F})(50 \text{ sq ft})(48°F) + (0.13 \text{ Btuh/sq ft-°F})(50 \text{ sq ft})(23°F) = 461.5 \text{ Btuh}$$

At hour 14:00 —

$$q = (0.13 \text{ Btuh/sq ft-°F})(50 \text{ sq ft})(42°F) + (0.13 \text{ Btuh/sq ft-°F})(50 \text{ sq ft})(37°F) = 513.5 \text{ Btuh}$$

At hour 16:00 —

$$q = (0.13 \text{ Btuh/sq ft-°F})(50 \text{ sq ft})(37°F) + (0.13 \text{ Btuh/sq ft-°F})(50 \text{ sq ft})(42°F) = 513.5 \text{ Btuh}$$

In this case, the peak load happens to occur twice, once at hour 14:00 and again at hour 16:00. The peak design exterior cooling load is 513.5, or 514 Btuh.

The above is a simplified example of a space with multiple heat gains and cooling loads. For an actual room, there are usually many heat gains with peak cooling loads occurring at different times. This is one of the reasons why it is necessary to use considerable judgment when using the CLTD-CLF method to calculate cooling loads. This will become increasingly evident as additional sources of heat gain are discussed in this and the next chapter.

It may also be observed in the above example that the same correction factors were used over again each hour for each wall. Since multiple calculations are frequently necessary, it is usually easier to calculate the correction factors for each surface separately, and then apply them to the hourly CLTD values.

For walls that have a U value greater than those listed in Table 7-1 for a given wall weight, the group selection must be adjusted. For each 7.0 increase in the R value (or a U value that is 14% lower) due to added insulation, use the previous alphabetic, or heavier, wall group letter. For example, if a given wall is similar in actual weight to a Group C wall but has insulation that is R-7 or greater, the calculations should be based on Group B CLTD values. If the given design wall is already similar to a Group A wall and has insulation that is R-7 or greater than that listed for a Group A wall, a CLTD should be used from Table 7-4.

N	NE	E	SE	S	SW	W	NW
11	17	22	21	17	21	22	17

Table 7-4. CLTD for Group A walls with additional insulation.

Cooling loads for sunlit roofs

The procedure for calculating cooling loads for sunlit roofs is similar to the procedure for sunlit walls. ASHRAE has grouped roofs by construction type or weight and also according to whether or not there is a suspended ceiling. The roof types, along with the weight of the representative U values and the cooling loads, are given in Table 7-5.

The same assumed conditions for the wall CLTDs apply to the roof CLTDs. Therefore, the roof CLTD values must also be corrected. In addition to the conditions assumed for the walls, the representative roof types have the following assumed conditions:

- The roof does not have a forced ventilation system, such as a fan in the space, or attic, between the ceiling and roof. The space is essentially a "dead air space."

- The roof is flat.

Although roofs are not flat in many cases, there is no correction for an inclined roof. Using the values for a flat roof should not introduce significant errors, since the flat roof is probably the worst-case situation. During summer months, the

Roof No	Description of Construction	Weight, lb/ft²	U-value, Btu/(h·ft²·°F)	1	2	3	4	5	6	7	8	9	10	11	12	13	14	15	16	17	18	19	20	21	22	23	24	Hour of Maximum CLTD	Minimum CLTD	Maximum CLTD	Difference CLTD
											Solar Time																				
	**Without Suspended Ceiling**																														
1	Steel sheet with 1-in. (or 2-in.) insulation	7 (8)	0.213 (0.124)	1	-2	-3	-3	-5	-3	6	19	34	49	61	71	78	79	77	70	59	45	30	18	12	8	5	3	14	-5	79	84
2	1-in. wood with 1-in. insulation	8	0.170	6	3	0	-1	-3	-3	-2	4	14	27	39	52	62	70	74	74	70	62	51	38	28	20	14	9	16	-3	74	77
3	4-in. lightweight concrete	18	0.213	9	5	2	0	-2	-3	-3	1	9	20	32	44	55	64	70	73	71	66	57	45	34	25	18	13	16	-3	73	76
4	2-in. heavyweight concrete with 1-in. (or 2-in.) insulation	29 (0.122)	0.206	12	8	5	3	0	-1	-1	3	11	20	30	41	51	59	65	66	66	62	54	45	36	29	22	17	16	-1	67	68
5	1-in. wood with 2-in. insulation	9	0.109	3	0	-3	-4	-5	-7	-6	-3	5	16	27	39	49	57	63	64	62	57	48	37	26	18	11	7	16	-7	64	71
6	6-in. lightweight concrete	24	0.158	22	17	13	9	6	3	1	1	3	7	15	23	33	43	51	58	62	64	62	57	50	42	35	28	18	1	64	63
7	2.5-in. wood with 1-in. ins.	13	0.130	29	24	20	16	13	10	7	6	6	9	13	20	27	34	42	48	53	55	56	54	49	44	39	34	19	6	56	50
8	8-in. lightweight concrete	31	0.126	35	30	26	22	18	14	11	9	7	7	9	13	19	25	33	39	46	50	53	54	53	49	45	40	20	7	54	47
9	4-in. heavyweight concrete with 1-in. (or 2-in.) insulation	52 (52)	0.200 (0.120)	25	22	18	15	12	9	8	8	10	14	20	26	33	40	46	50	53	53	52	48	43	38	34	30	18	8	53	45
10	2.5-in. wood with 2-in. ins.	13	0.093	30	26	23	19	16	13	10	9	8	9	13	17	23	29	36	41	46	49	51	50	47	43	39	35	19	8	51	43
11	Roof terrace system	75	0.106	34	31	28	25	22	19	16	14	13	13	15	18	22	26	31	36	40	44	45	46	45	43	40	37	20	13	46	33
12	6-in. heavyweight concrete with 1-in. (or 2-in.) insulation	75 (75)	0.192 (0.117)	31	28	25	22	20	17	15	14	14	16	18	22	26	31	36	40	43	45	45	44	42	40	37	34	19	14	45	31
13	4-in. wood with 1-in. (or 2-in.) insulation	17 (18)	0.106 (0.078)	38	36	33	30	28	25	22	20	18	17	16	17	18	21	24	28	32	36	39	41	43	43	42	40	22	16	43	27
	**With Suspended Ceiling**																														
1	Steel Sheet with 1-in. (or 2-in.) insulation	9 (10)	0.134 (0.092)	2	0	-2	-3	-4	-4	-1	9	23	37	50	62	71	77	78	74	67	56	42	28	18	12	8	5	15	-4	78	82
2	1-in. wood with 1-in. ins.	10	0.115	20	15	11	8	5	3	2	3	7	13	21	30	40	48	55	60	62	61	58	51	44	37	30	25	17	2	62	60
3	4-in. lightweight concrete	20	0.134	19	14	10	7	4	2	0	0	4	10	19	29	39	48	56	62	65	64	61	54	46	38	30	24	17	0	65	65
4	2-in. heavyweight concrete with 1-in. insulation	30	0.131	28	25	23	20	17	15	13	13	14	16	20	25	30	35	39	43	46	47	46	44	41	38	35	32	18	13	47	34
5	1-in. wood with 1-in. ins.	10	0.083	25	20	16	13	10	7	5	5	7	12	18	25	33	41	48	53	57	57	56	52	46	40	34	29	18	5	57	52
6	6-in. lightweight concrete	26	0.109	32	28	23	19	16	13	10	8	7	8	11	16	22	29	36	42	48	52	54	54	51	47	42	37	20	7	54	47
7	2.5-in. wood with 1-in. insulation	15	0.096	34	31	29	26	23	21	18	16	15	15	16	18	21	25	30	34	38	41	43	44	44	42	40	37	21	15	44	29
8	8-in. lightweight concrete	33	0.093	39	36	33	29	26	23	20	18	15	14	14	15	17	20	25	29	34	38	42	45	46	45	44	42	21	14	46	32
9	4-in. heavyweight concrete with 1-in. (or 2-in.) ins.	53 (54)	0.128 (0.090)	30	29	27	26	24	22	21	20	20	21	22	24	27	29	32	34	36	38	38	38	37	36	34	33	19	20	38	18
10	2.5-in. wood with 2-in. ins.	15	0.072	35	33	30	28	26	24	22	20	18	18	18	20	22	25	28	32	35	38	40	41	41	40	39	37	21	18	41	23
11	Roof terrace system	77	0.082	30	29	28	27	26	25	24	23	22	22	22	23	23	25	26	28	29	31	32	33	33	33	33	32	22	22	33	11
12	6-in. heavyweight concrete with 1-in. (or 2-in.) insulation	77 (77)	0.125 (0.088)	29	28	27	26	25	24	23	22	21	21	22	23	25	26	28	30	32	33	34	34	34	33	32	31	20	21	34	13
13	4-in. wood with 1-in. (or 2-in.) insulation	19 (20)	0.082 (0.064)	35	34	33	32	31	29	27	26	24	23	22	21	22	22	24	25	27	30	32	34	35	36	37	36	23	21	37	16

(1) *Direct Application of Table Without Adjustments:*

Values were calculated using the following conditions:

- Dark flat surface roof ("dark" for solar radiation absorption)
- Indoor temperature of 78 °F
- Outdoor maximum temperature of 95 °F with outdoor mean temperature of 85 °F and an outdoor daily range of 21 °F
- Solar radiation typical of 40 deg North latitude on July 21
- Outside surface resistance, $R_o = 0.333$ ft²·°F·h/Btu
- Without and with suspended ceiling, but no attic fans or return air ducts in suspended ceiling space
- Inside surface resistance, $R_i = 0.685$ ft²·°F·h/Btu

(2) *Adjustments to Table Values:*

The following equation makes adjustments for deviations of design and solar conditions from those listed in (1) above.

$$CLTD_{corr} = [(CLTD + LM) K + (78 - t_R) + (t_o - 85)] f$$

where CLTD is from this table

(a) LM is latitude-month correction from Table for a horizontal surface.

(b) K is a color adjustment factor applied after first making month-latitude adjustments. Credit should not be taken for a light-colored roof except where permanence of light color is established by experience, as in rural areas or where there is little smoke.

K = 1.0 if dark colored or light in an industrial area
K = 0.5 if permanently light-colored (rural area)

(c) $(78 - t_R)$ is Indoor design temperature correction

(d) $(t_o - 85)$ is outdoor design temperature correction, where t_o is the average outside temperature on design day

(e) f is a factor for attic fan and or ducts above ceiling applied after all other adjustments have been made

f = 1.0 no attic or ducts
f = 0.75 positive ventilation

Values in Table were calculated without and with suspended ceiling, but make no allowances for positive ventilation or return ducts through the space. If ceiling is insulated and fan is used between ceiling and roof, CLTD may be reduced 25% (f = 0.75). Analyze use of the suspended ceiling space for a return air plenum or with return air ducts separately.

(3) *Roof Constructions Not Listed in Table:*

The U-Values listed are only guides. The actual value of U as obtained from tables such as Table : or as calculated for the actual roof construction should be used.

An actual roof construction not in this table would be thermally similar to a roof in the table, if it has similar mass, lb/ft², and similar heat capacity, Btu/ft²·°F. In this case, use the CLTD from this table as corrected

Example: A flat roof without suspended ceiling has mass = 18.0 lb/ft², U = 0.20 Btu/h·ft²·°F, and heat capacity = 9.5 Btu/ft²·°F.

Use $CLTD_{uncorr}$ from Roof No. 13, to obtain $CLTD_{corr}$ and use the actual U value to calculate $q/A = U (CLTD_{corr}) = 0.20 (CLTD_{corr})$.

(4) *Additional Insulation:*

For each R-7 increase in R-value from insulation added to the roof structure, use a CLTD for a roof whose weight and heat capacity are approximately the same, but whose CLTD has a maximum value 2 h later. If this is not possible, because a roof with longest time lag has already been selected, use an effective CLTD in cooling load calculation equal to 29 °F.

Table 7-5. Cooling load temperature differences for flat roofs. (Copyright 1989 by the American Society of Heating, Refrigerating and Air-Conditioning Engineers, Inc., from the ASHRAE *Handbook—Fundamentals*. Used by permission.)

sun is almost directly overhead and the solar radiation would fall directly on a flat roof.

The equation correcting the CLTD values listed is identical to Equation 7.2 for sunlit walls, except for the addition of an attic ventilation factor. The following equation may be used to correct the CLTD values obtained from Table 7-5:

$$CLTD_c = [(CLTD + LM)(k) + (78°F - T_i) + (T_{oo} - 85°F)](f)$$

Equation 7.3

where:

f = factor for attic ventilation after all other adjustments have been made. ($f = 1.0$ for no ventilation; 0.75 for positive ventilation)

All other correction variables in Equation 7.3 are the same as for Equation 7.2. (Note that when using Table 7-3 to correct for the latitude and month, the column headed HOR should be used for roofs.)

As in the case of walls, the U value in Table 7-5 is for comparison only. The actual U value of the roof for the building being designed should be used in Equation 7.1

Example 7-3.

Assume that a dark 200-sq ft flat roof similar to the one described in Example 4-3 is on a building in Chicago on August 21. The weight of the roof has been calculated to be 8.72 lb/sq ft, with a U value of 0.09 Btuh/sq ft-°F. The attic air space has no positive ventilation, and a room below is to be maintained at 75°F. What is the peak cooling load for the roof, and at what hour does it occur?

From Table 7-5, a Roof Number 1 with a suspended ceiling would be chosen as the closest matching roof. The peak CLTD for the representative roof is 78°F, which occurs at hour 15:00.

From Table 3-1, the 2-1/2% climatic design conditions for Chicago are 91°F dry bulb/74°F

wet bulb. The daily temperature range is 20°F. Chicago's latitude is 42-degrees north.

Therefore, the average daily dry bulb temperature is:

$$T_{oo} = 91°F - \frac{20°F}{2} = 81°F$$

Using Equation 7.3 to correct the CLTD:

$$CLTD_c = [(78°F + -3)(1) + (78°F - 75°F) + (81°F - 85°F)](1) = 74°F$$

Using Equation 7.1 and the calculated U value for the roof, the peak cooling load from the roof is:

$$q = (0.09 \text{ Btuh/sq ft-°F})(200 \text{ sq ft})(74°F) = 1,332 \text{ Btuh}$$
at hour 15:00

In the above example, the air space between the roof deck and the suspended ceiling was a dead air space, with little or no ventilation or air movement, as was the case in Example 4-3. For spaces with ceiling return-air plenums, the roof cooling load calculations should be based on the roof classification without a suspended ceiling.

Not all of the roof cooling load becomes a cooling load for the room below the ceiling, though. Much of the cooling load becomes a return-air cooling load for the hvac system, but not a load for the room. This is an important distinction; the air supplied to the room should be based on the actual calculated room cooling load.

Heat gains for windows and glass

In the previous sections of this chapter, the effect of solar radiation on opaque surfaces, such as walls and roofs, was discussed. The resulting effect on the cooling load was also discussed. The effect of solar radiation on glass areas, or fenestration areas, can have an even more significant effect on the cooling load.

The space cooling load that results from solar radiation on fenestration areas depends upon a number of variables, including:

- the size of the fenestration area;

- the direction the fenestration is facing, or its orientation;

- exterior and interior shading;

- solar radiation intensity and angle of incidence;

- type of glazing material and number of layers;

- time of day and day of the year;

- mass of the space.

Any solar radiation falling on a surface, such as glass, will either be reflected, absorbed, or transmitted through the glass, as was shown in Figure 2-2. The relative proportions of reflected, absorbed, and transmitted energy are primarily a function of the glazing material and the angle of incidence of the radiation. Figure 7-2 shows typical proportions of reflected, absorbed, and transmitted energy for sunlit glass.

The total amount and intensity of solar radiation incident on a surface, such as a window, is a function of the angle of incidence at which the radiation strikes the glazing surface and the extent to which the surface is shaded by any external objects. The angle of incidence is a function of the surface orientation, location, time of day, and day of the year.

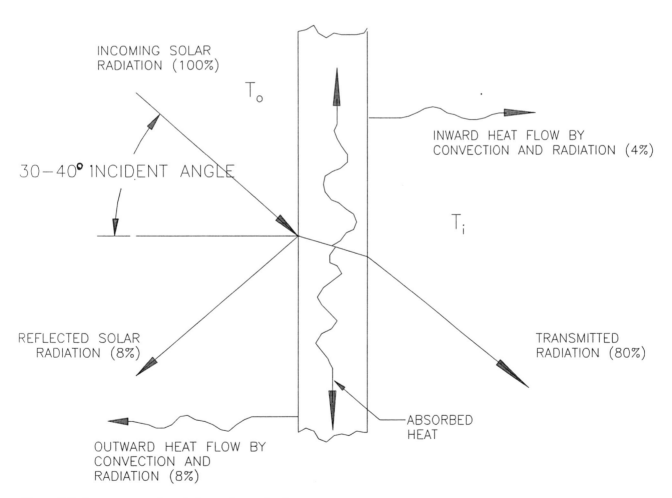

Figure 7-2. Instantaneous heat balance for sunlit glass.

Angle of incidence for solar radiation

The earth rotates on a nearly north-south axis every 24 hours. It also moves in a slightly elliptical orbit around the sun, called the elliptical plane. As a result, the incident solar radiation and the angle of incidence of that radiation is constantly changing. These changes in intensity and angle have a significant effect on the resultant solar heat gain for surfaces.

Because the sun is so far away from the earth, it may be considered a point source of radiation. Thus, the relationship of the sun with respect to a surface may be described by angles and trigonometric relationships. The primary angles that determine the angle of incidence of solar radiation on any given surface are the solar altitude of the sun, the solar azimuth of the sun, the surface-solar azimuth, and the surface tilt angle. These angles are shown in Figures 7-3 and 7-4.

For a vertical surface, the surface-solar azimuth (γ) is the angle measured in the horizontal plane between the projection of the sun's rays on that plane and a line perpendicular to the surface. The angle can be positive or negative, with a surface directly facing south having a $\gamma = 0$. The solar azimuth angle (ϕ) is the angle in the horizontal plane measured between south and the projection of the sun's rays on that plane. The angle can be positive or negative, with ϕ being positive for west of south, or afternoon. When the sun is due south of a surface, $\phi = 0$.

The solar altitude (the sun's altitude angle, β) is the angle between the sun's rays and the projection of a ray on a horizontal surface. It is simply the angle of the sun above the horizon and can vary from $\beta = 0$ degree to $\beta = 90$ degrees. It is particularly useful when calculating the effect of external shading. The angle of tilt (α) is the angle between the surface and the horizontal plane.

The solar declination (δ) is the angle between the sun's rays and the earth's equatorial plane. It is similar to the solar altitude, except that the angle is measured between the sun's rays and the equatorial plane instead of the horizontal plane.

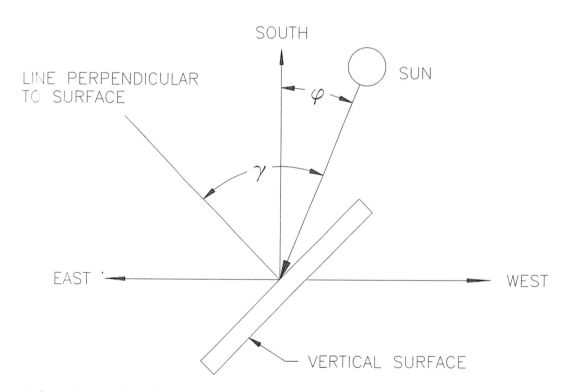

Figure 7-3. Surface-solar azimuth and solar azimuth.

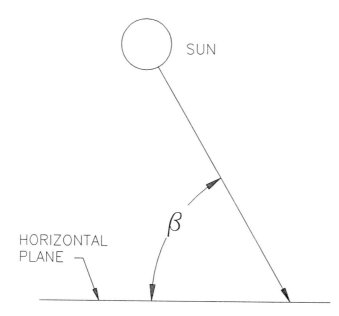

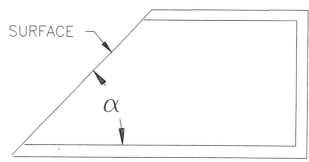

Figure 7-4. Solar altitude and surface tilt angle.

The solar declination may be approximated using the following equation:

$$\delta = 23.45 \sin\left[(360°)\left(\frac{284 + n}{365}\right)\right]$$

Equation 7.4

where:

n = day of the year

Values of solar declination, at mid-month, are as follows:

January	-21.4
February	-14.0
March	- 2.8
April	9.1
May	18.6
June	23.1
July	21.4
August	14.0
September	2.8
October	- 9.1
November	-18.6
December	-23.1

The solar altitude may be determined by using the following equation:

$$\beta = \sin^{-1}(\sin \delta \sin L + \cos \delta \cos L \cos \varpi)$$

Equation 7.5

where:

L = latitude, with north being positive
δ = angle of solar declination
ϖ = hour angle, i.e., the number of hours from solar noon times 15 degrees, with mornings being positive

Glazing materials

Heat gain through glazing materials, such as glass, occurs by two methods. The primary method of heat transfer through glass is transmitted solar radiation. The secondary method is by conduction through the glass. Conduction heat transfer and radiation heat transfer were discussed in Chapter 2.

Glazing materials can include clear and opaque plastics as well as glass. The amount of solar radiation transmitted by a glazing material depends on the solar optical properties, which include the wavelength of radiation, chemical composition of the material, coloring in the glazing material, and angle of incidence of the radiation.

One measure of a particular glazing material's ability to transmit solar radiation is the index of refraction. The index of refraction measures how much the path of light passing through a material is changed. The higher the index of refraction, the more a beam of light changes direction as it passes through the material, and the more it is reflected by the material.

An alternate method used to determine the ability of a particular glazing, or combination of glazing, to transmit solar radiation is the shading coefficient (SC). The SC is defined as the ratio of solar gain through a given glazing system under a specific set of conditions, to the solar gain through a reference glass under the same conditions. Use the following equation to determine the SC:

$$SC = \frac{\text{Solar heat gain of fenestration}}{\text{Solar heat gain of reference glass}}$$

Equation 7.6

The reference glass is a single sheet of unshaded, clear, double-strength glass (DSA) with a transmittance of 0.86, reflectance of 0.08, and absorbance of 0.06.

Typical SCs are given in Table 7-6. The SC for any fenestration will increase the tabulated values when the inner surface film coefficient is increased and the outer coefficient is decreased. The reverse is also true. Table 7-6 gives SCs for outside film coefficients of 3 and 4 Btuh/sq ft-°F. The larger film coefficient is for an outdoor wind speed of 7-1/2 mph, while the other is for lower wind speeds. The film convection coefficient affects the SC by affecting the amount of radiation absorbed by the glazing and then transferred away by convection.

The values presented in Table 7-6 are typical. Whenever possible, the glazing manufacturer's literature should be consulted to get actual values for SCs and U values.

A. Single Glass				
Type of Glass	Nominal Thickness[b]	Solar Trans.[b]	Shading Coefficient	
			h_0=4.0	h_0=3.0
Clear	1/8 in.	0.86	1.00	1.00
	1/4 in.	0.78	0.94	0.95
	3/8 in.	0.72	0.90	0.92
	1/2 in.	0.67	0.87	0.88
Heat Absorbing	1/8 in.	0.64	0.83	0.85
	1/4 in.	0.46	0.69	0.73
	3/8 in.	0.33	0.60	0.64
	1/2 in.	0.24	0.53	0.58
B. Insulating Glass				
Clear Out, Clear In	1/8 in.[c]	0.71[e]	0.88	0.88
Clear Out, Clear In	1/4 in.	0.61	0.81	0.82
Heat Absorbing[d] Out, Clear In	1/4 in.	0.36	0.55	0.58

[a]Refers to factory-fabricated units with 3/16, 1/4, or 1/2-in. air space or to prime windows plus storm sash.
[b]Refer to manufacturer's literature for values.
[c]Thickness of each pane of glass, not thickness of assembled unit.
[d]Refers to gray, bronze, and green tinted heat-absorbing float glass.
[e]Combined transmittance for assembled unit.

Table 7-6. Shading coefficients for single glass and insulating glass. (Copyright 1989 by the American Society of Heating, Refrigerating and Air-Conditioning Engineers, Inc., from the ASHRAE Handbook—Fundamentals Used by permission.)

Shading

The amount of solar heat gain to a space through fenestration areas is also directly related to the extent to which fenestration areas are shaded, both internally and externally.

External shading occurs when an object outside the building partially or fully blocks the sun's rays and shades the fenestration area. Such objects include overhangs, side fins, other buildings or objects, and vegetation (such as trees). External shading has a much greater effect on space heat gains and cooling loads than interior shading. It is possible for an object to shade itself. For a vertical surface, such as a wall, whenever the surface-solar azimuth angle is greater than 90 degrees and less than 180 degrees, the surface will be shaded by itself. In effect, the surface "has its back to the sun." Surfaces that face directly north will always be shaded.

The most common form of external shading is the overhang. The amount, or fraction, of the fenestration surface that is shaded may be estimated by the shading factor S. The shading factor is defined as the fraction of glazing area shaded by an external object. The shading factor for an overhang may be estimated by applying the following equation:

$$S = \left(\frac{1}{H} \right)(W \tan \beta - D)$$

Equation 7.7

where:

W = width of the overhang as it extends out from the surface

D = vertical distance from the top of the fenestration area to the underside of the overhang

H = height of the glazing

β = solar altitude

Figure 7-5 graphically shows the variables for Equation 7.7.

If S is equal to or less than 0, all of the fenestration area is in the sun. If S is equal to or greater than 1.0, the fenestration area is completely shaded. Any value of S between 0 and 1.0 is the fraction of the glazing that is shaded.

Equation 7.7 is only a good approximation for estimating the fraction of the glazing that is shaded. It does not take the length of the overhang or the fenestration into account. For more complex overhangs, as well as side fins, the reader is referred to Chapter 3 of Reference 1.

Interior shading, such as blinds, drapes, or shades, is frequently employed for solar control. Interior shading has two effects. First, it reflects much of the solar radiation at the fenestration opening so the direct radiation never enters the space. Second, the interior shade prevents solar

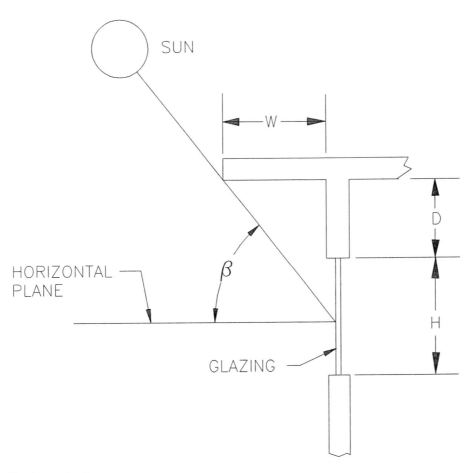

Figure 7-5. Overhang shading.

radiation from being absorbed by the interior mass of the space itself. Instead, the interior shading absorbs much of the radiation which, in turn, is convected to the air inside the space.

Since the solar radiation is absorbed by the interior shading and then convected to the room air, interior shades tend to be less effective in reducing solar heat gain than exterior shades. Table 7-7 gives typical SCs for fenestration areas with venetian blinds and roller shades.

Shading coefficients of draperies for single- and multi-layer glazing are complex results of the color and weave of the drapery fabric. Whenever possible, SCs for draperies should be obtained from the drapery or fabric manufacturer. If manufacturers' data is not readily available, refer to Chapter 27 of Reference 1 or Chapter 3 of Reference 4.

Cooling loads for windows and glass

Heat gain through a fenestration area to a space occurs by two methods: solar radiation and conduction heat transfer. Conduction heat transfer occurs as the result of the difference in temperature between indoors and outdoors and radiant heat that has been absorbed by the glazing material itself. Equation 7.8 shows the components that make up the total heat flow through glass:

$$q_t = q_r + q_a = q_c$$

Equation 7.8

where:

q_t = total or net heat transmission through the glass (Btuh)

q_r = radiation heat transfer (Btuh)

q_a = inward flow of absorbed radiation (Btuh)

q_c = heat conducted through the glazing material (Btuh)

	Nominal Thickness[a], Inches	Solar Trans.[b]	Type of Shading				
			Venetian Blinds		Roller Shade		
					Opaque		Translucent
			Medium	Light	Dark	White	Light
Clear	3/32[c]	0.87 to 0.79	0.74[d] (0.63)[e]	0.67[d] (0.58)[e]	0.81	0.39	0.44
Clear	1/4 to 1/2	0.80 to 0.71					
Clear Pattern	1/8 to 1/2	0.87 to 0.79					
Heat-Absorbing Pattern	1/8	—					
Tinted	3/16, 7/32	0.74, 0.71					
Heat-Absorbing[f]	3/16, 1/4	0.46					
Heat-Absorbing Pattern	3/16, 1/4	—	0.57	0.53	0.45	0.30	0.36
Tinted	1/8, 7/32	0.59, 0.45					
Heat-Absorbing or Pattern	—	0.44 to 0.30	0.54	0.52	0.40	0.28	0.32
Heat-Absorbing[f]	3/8	0.34					
Heat-Absorbing or Pattern	—	0.29 to 0.15 0.24	0.42	0.40	0.36	0.28	0.31
Reflective Coated Glass	SC[g] = 0.30		0.25	0.23			
	= 0.40		0.33	0.29			
	= 0.50		0.42	0.38			
	= 0.60		0.50	0.44			

[a]Refer to manufacturer's literature for values.
[b]For vertical blinds with opaque white and beige louvers in the tightly closed position, SC is 0.25 and 0.29 when used with glass of 0.71 to 0.80 transmittance.
[c]Typical residential glass thickness
[d]From Van Dyck and Konen (1982), for 45° open venetian blinds, 35° solar incidence, and 35° profile angle.
[e]Values for closed venetian blinds. Use these values only when operation is automated for solar gain reduction (as opposed to daylight use).
[f]Refers to gray, bronze, and green tinted heat-absorbing glass.
[g]SC for glass with no shading device.

Table 7-7. Shading coefficients for single glass with indoor shading by venetian blinds or roller shades. (Copyright 1989 by the American Society of Heating, Refrigerating and Air-Conditioning Engineers, Inc., from the ASHRAE Handbook—Fundamentals. Used by permission.)

Note that q_a and q_c may be positive or negative (i.e., the flow of heat may be into or out of the space).

In a manner similar to cooling loads for sunlit walls and roofs, ASHRAE has developed CLTDs and Cooling Load Factors (CLFs) to account for transient heat transfer and thermal storage of building materials.

The conduction heat gain and resultant cooling load is the sum of the absorbed radiant heat and the conducted heat due to the difference between indoor and outdoor temperatures. Equation 7.1 may be used to calculate the cooling load due to conduction through fenestration areas. Table 7-8 provides CLTDs for conduction through glass, to be used in Equation 7.1.

The CLTD values listed in Table 7-8 are valid for an indoor temperature of 78°F, an outdoor temperature between 93° and 102°F, and an outdoor daily range between 16° and 34°F, with an average daily temperature of 85°F. For conditions other than those listed above, CLTD values must be corrected in accordance with the footnote provided with Table 7-8.

Referring again to Equation 7.8, the total heat transmission through a fenestration area also includes radiation heat transfer. Again, ASHRAE has developed a method to account for thermal storage by the building mass. For solar radiation, CLFs are used.

CLFs are different from CLTDs in that they account for heat gains that are inside or have directly entered the space and are absorbed within the space itself. Such is the case for solar radiation that shines in through a window, is absorbed by floors, walls, and objects, and is then released later as a cooling load on the space. The magnitude and time of the peak space cooling load are functions of the thermal mass of the space.

Equation 7.9 may be used to calculate the radiant cooling load on a space due to solar radiation through fenestration areas:

$$q_r = (A)(SC)\ (SHGF)\ (CLF)$$

Equation 7.9

where:
A = net glazing area of the fenestration area (sq ft)
SC = shading coefficient
$SHGF$ = maximum solar heat gain factor (Btu/sq ft) (from Table 7-9)
CLF = cooling load factor (from Tables 7-10, 7-11, or 7-12 for the appropriate room characteristics)

Referring to Table 7-9, the values listed are maximum solar heat gain for various fenestration orientations at various north latitudes for each month. These values are typical values that take the solar angle of incidence into account.

Whenever a fenestration area is fully or partially shaded, the maximum solar heat gain and CLF for north-facing fenestration should be applied to the shaded portion. This accounts for radiant heat gain due to indirect or reflected radiation. Note also that the CLF values in Table 7-10 are for glass without internal shading and without carpeting on the floors.

Solar time, h	CLTD °F	Solar time, h	CLTD °F
0100	1	1300	12
0200	0	1400	13
0300	-1	1500	14
0400	-2	1600	14
0500	-2	1700	13
0600	-2	1800	12
0700	-2	1900	10
0800	0	2000	8
0900	2	2100	6
1000	4	2200	4
1100	7	2300	3
1200	9	2400	2

Corrections: The values in the table were calculated for an inside temperature of 78°F and an outdoor maximum temperature of 95°F with an outdoor daily range of 21°F. The table remains approximately correct for other outdoor maximums 93 to 102°F and other outdoor daily ranges 16 to 34°F, provided the outdoor daily average temperature remains approximately 85°F. If the room air temperature is different from 78°F and/or the outdoor daily average temperature is different from 85°F, the following rules apply: (a) For room air temperature less than 78°F, add the difference between 78°F and room air temperature; if greater than 78°F, subtract the difference. (b) For outdoor daily average temperature less than 85°F, subtract the difference between 85°F and the daily average temperature; if greater than 85°F, add the difference.

Table 7-8. CLTDs for glass. (Copyright 1989 by the American Society of Heating, Refrigerating and Air-Conditioning Engineers, Inc., from the ASHRAE *Handbook—Fundamentals.* Used by permission.)

0° N Lat

	N	NNE/NNW	NE/NW	ENE/WNW	E/W	ESE/WSW	SE/SW	SSE/SSW	S	HOR
Jan.	34	34	88	177	234	254	235	182	118	296
Feb.	36	39	132	205	245	247	210	141	67	306
Mar.	38	87	170	223	242	223	170	87	38	303
Apr.	71	134	193	224	221	184	118	38	37	284
May	113	164	203	218	201	154	80	37	37	265
June	129	173	206	212	191	140	66	37	37	255
July	115	164	201	213	195	149	77	38	38	260
Aug.	75	134	187	216	212	175	112	39	38	276
Sept.	40	84	163	213	231	213	163	84	40	293
Oct.	37	40	129	199	236	238	202	135	66	299
Nov.	35	35	88	175	230	250	230	179	117	293
Dec.	34	34	71	164	226	253	240	196	138	288

4° N Lat

	N	NNE/NNW	NE/NW	ENE/WNW	E/W	ESE/WSW	SE/SW	SSE/SSW	S	HOR
Jan.	33	33	79	170	229	252	237	193	141	286
Feb.	35	35	123	199	242	248	215	152	88	301
Mar.	38	77	163	219	242	227	177	96	43	302
Apr.	55	125	189	223	223	190	126	43	38	287
May	93	154	200	220	206	161	89	38	38	272
June	110	164	202	215	196	147	73	38	38	263
July	96	154	197	215	200	156	85	39	38	267
Aug.	59	124	184	215	214	181	120	42	40	279
Sep.	39	75	156	209	231	216	170	93	44	293
Oct.	36	36	120	193	234	239	207	148	86	294
Nov.	34	34	79	168	226	248	232	190	139	284
Dec.	33	33	62	157	221	250	242	206	160	277

8° N Lat

	N	NNE/NNW	NE/NW	ENE/WNW	E/W	ESE/WSW	SE/SW	SSE/SSW	S	HOR
Jan.	32	32	71	163	224	250	242	203	162	275
Feb.	34	34	114	193	239	248	219	165	110	294
Mar.	37	67	156	215	241	230	184	110	55	300
Apr.	44	117	184	221	225	195	134	53	39	289
May	74	146	198	220	209	167	97	39	38	277
June	90	155	200	217	200	141	82	39	39	269
July	77	145	195	215	204	162	93	40	39	272
Aug.	47	117	179	214	216	186	128	51	41	282
Sep.	38	66	149	205	230	219	176	107	56	290
Oct.	35	35	112	187	231	239	211	160	108	288
Nov.	33	33	71	161	220	245	233	200	160	273
Dec.	31	31	55	149	215	246	247	215	179	265

12° N Lat

	N	NNE/NNW	NE/NW	ENE/WNW	E/W	ESE/WSW	SE/SW	SSE/SSW	S	HOR
Jan.	31	31	63	155	217	246	247	212	182	262
Feb.	34	34	105	186	235	248	226	177	133	286
Mar.	36	58	148	210	240	233	190	124	73	297
Apr.	40	108	178	219	227	200	142	64	40	290
May	60	139	194	220	212	173	106	40	40	260
June	75	149	198	217	204	161	90	40	40	274
July	63	139	191	215	207	168	102	41	41	275
Aug.	42	109	174	212	218	191	135	62	142	282
Sep.	37	57	142	201	229	222	182	121	73	287
Oct.	34	34	103	180	227	238	219	172	130	280
Nov.	32	32	63	153	214	241	243	209	179	260
Dec.	30	30	47	141	207	242	251	223	197	250

16° N Lat

	N	NNE/NNW	NE/NW	ENE/WNW	E/W	ESE/WSW	SE/SW	SSE/SSW	S	HOR
Jan.	30	30	55	147	210	244	251	223	199	248
Feb.	33	33	96	180	231	247	233	188	154	275
Mar.	35	53	140	205	239	235	197	138	93	291
Apr.	39	99	172	215	227	204	150	77	45	289
May	52	132	189	218	215	179	115	45	41	282
June	66	142	194	217	207	167	99	41	41	277
July	55	132	187	214	210	174	111	44	42	277
Aug.	41	100	168	209	219	196	143	74	46	282
Sep.	36	50	134	196	227	224	191	134	93	282
Oct.	33	33	95	174	223	237	225	183	150	270
Nov.	30	30	55	145	206	241	247	220	196	246
Dec.	29	29	41	132	198	241	254	233	212	234

20° N Lat

	N	NNE/NNW	NE/NW	ENE/WNW	E/W	ESE/WSW	SE/SW	SSE/SSW	S	HOR
Jan.	29	29	48	138	201	243	253	233	214	232
Feb.	31	31	88	173	226	244	238	201	174	263
Mar.	34	49	132	200	237	236	206	152	115	284
Apr.	38	92	166	213	228	208	155	91	58	287
May	47	123	184	217	217	184	124	54	42	283
June	59	135	189	216	210	173	108	45	42	279
July	48	124	182	213	212	179	119	53	43	278
Aug.	40	91	162	206	220	200	152	88	57	280
Sep.	36	46	127	191	225	225	199	148	114	275
Oct.	32	32	87	167	221	236	231	196	170	258
Nov.	29	29	48	136	197	239	249	229	211	230
Dec.	27	27	35	122	187	238	254	241	226	217

24° N. Lat

	N	NNE/NNW	NE/NW	ENE/WNW	E/W	ESE/WSW	SE/SW	SSE/SSW	S	HOR
Jan.	27	27	41	128	190	240	253	241	227	214
Feb.	30	30	80	165	220	244	243	213	192	249
Mar.	34	45	124	195	234	237	214	168	137	275
Apr.	37	88	159	209	228	212	169	107	75	283
May	43	117	178	214	218	190	132	67	46	282
June	55	127	184	214	212	179	117	55	43	279
July	45	116	176	210	213	185	129	65	46	278
Aug.	38	87	156	203	220	204	162	103	72	277
Sep.	35	42	119	185	222	225	206	163	134	266
Oct.	31	31	79	159	211	237	235	207	187	244
Nov.	27	27	42	126	187	236	249	237	224	213
Dec.	26	26	29	112	180	234	247	247	237	199

28° N. Lat

	N (Shade)	NNE/NNW	NE/NW	ENE/WNW	E/W	ESE/WSW	SE/SW	SSE/SSW	S	HOR
Jan.	25	25	35	117	183	235	251	247	238	196
Feb.	29	29	72	157	213	244	246	224	207	234
Mar.	33	41	116	189	231	237	221	182	157	265
Apr.	36	84	151	205	228	216	178	124	94	278
May	40	115	172	211	219	195	144	83	58	280
June	51	125	178	211	213	184	128	68	49	278
July	41	114	170	208	215	190	140	80	57	276
Aug.	38	83	149	199	220	207	172	120	91	272
Sep.	34	38	111	179	219	226	213	177	154	256
Oct.	30	30	71	151	204	236	238	217	202	229
Nov.	26	26	35	115	181	232	247	243	235	195
Dec.	24	24	24	99	172	227	248	251	246	179

32° N. Lat

	N (Shade)	NNE/NW	NE/NW	ENE/WNW	E/W	ESE/WSW	SE/SW	SSE/SSW	S	HOR
Jan.	24	24	29	105	175	229	249	250	246	176
Feb.	27	27	65	149	205	242	248	232	221	217
Mar.	32	37	107	183	227	237	227	195	176	252
Apr.	36	80	146	200	227	219	187	141	115	271
May	38	111	170	208	220	199	155	99	74	276
June	44	122	176	208	214	189	139	83	60	276
July	40	111	167	204	215	194	150	96	72	273
Aug.	37	79	141	195	219	210	181	136	111	265
Sept.	33	35	103	173	215	227	218	189	171	244
Oct.	28	28	63	143	195	234	239	225	215	213
Nov.	24	24	29	103	173	225	245	246	243	175
Dec.	22	22	22	84	162	218	246	252	252	158

36° N. Lat

	N (Shade)	NNE/NNW	NE/NW	ENE/WNW	E/W	ESE/WSW	SE/SW	SSE/SSW	S	HOR
Jan.	22	22	24	90	166	219	247	252	252	155
Feb.	26	26	57	139	195	239	248	239	232	199
Mar.	30	33	99	176	223	238	232	206	192	238
Apr.	35	76	144	196	225	221	196	156	135	262
May	38	107	168	204	220	204	165	116	93	272
June	47	118	175	205	215	194	150	99	77	273
July	39	107	165	201	216	199	161	113	90	268
Aug.	36	75	138	190	218	212	189	151	131	257
Sep.	31	31	95	167	210	228	223	200	187	230
Oct.	27	27	56	133	187	230	239	231	225	195
Nov.	22	22	24	87	163	215	243	248	248	154
Dec.	20	20	20	69	151	204	241	253	254	136

Table 7-9. Maximum solar heat gain factor for sunlit glass Btu/sq ft. (Copyright 1989 by the American Society of Heating, Refrigerating and Air-Conditioning Engineers, Inc., from the ASHRAE Handbook—Fundamentals. Used by permission.)

40° N. Lat

	N (Shade)	NNE/NNW	NE/NW	ENE/WNW	E/W	ESE/WSW	SE/SW	SSE/SSW	S	HOR
Jan.	20	20	20	74	154	205	241	252	254	133
Feb.	24	24	50	129	186	234	246	244	241	180
Mar.	29	29	93	169	218	238	236	216	206	223
Apr.	34	71	140	190	224	223	203	170	154	252
May	37	102	165	202	220	208	175	133	113	265
June	48	113	172	205	216	199	161	116	95	267
July	38	102	163	198	216	203	170	129	109	262
Aug.	35	71	135	185	216	214	196	165	149	247
Sep.	30	30	87	160	203	227	226	209	200	215
Oct.	25	25	49	123	180	225	238	236	234	177
Nov.	20	20	20	73	151	201	237	248	250	132
Dec.	18	18	18	60	135	188	232	249	253	113

60° N. Lat

	N (Shade)	NNE/NNW	NE/NW	ENE/WNW	E/W	ESE/WSW	SE/SW	SSE/SSW	S	HOR
Jan.	7	7	7	7	46	88	130	152	164	21
Feb.	13	13	13	58	118	168	204	225	231	68
Mar.	20	20	56	125	173	215	234	241	242	128
Apr.	27	59	118	168	206	222	225	220	218	178
May	43	98	149	192	212	220	211	198	194	208
June	58	110	162	197	213	215	202	186	181	217
July	44	97	147	189	208	215	206	193	190	207
Aug.	28	57	114	161	199	214	217	213	211	176
Sep.	21	21	50	115	160	202	222	229	231	123
Oct.	14	14	14	56	111	159	193	215	221	67
Nov.	7	7	7	7	45	86	127	148	160	22
Dec.	4	4	4	4	16	51	76	100	107	9

44° N. Lat

	N (Shade)	NNE/NNW	NE/NW	ENE/WNW	E/W	ESE/WSW	SE/SW	SSE/SSW	S	HOR
Jan.	17	17	17	64	138	189	232	248	252	109
Feb.	22	22	43	117	178	227	246	248	247	160
Mar.	27	27	87	162	211	236	238	224	218	206
Apr.	33	66	136	185	221	224	210	183	171	240
May	36	96	162	201	219	211	183	148	132	257
June	47	108	169	205	215	203	171	132	115	261
July	37	96	159	198	215	206	179	144	128	254
Aug.	34	66	132	180	214	215	202	177	165	236
Sep.	28	28	80	152	198	226	227	216	211	199
Oct.	23	23	42	111	171	217	237	240	239	157
Nov.	18	18	18	64	135	186	227	244	248	109
Dec.	15	15	15	49	115	175	217	240	246	89

64° N. Lat

	N (Shade)	NNE/NNW	NE/NW	ENE/WNW	E/W	ESE/WSW	SE/SW	SSE/SSW	S	HOR
Jan.	3	3	3	3	15		67	89	96	8
Feb.	11	11	11	43	89	144	177	202	210	45
Mar.	18	18	47	113	159	203	226	236	239	105
Apr.	25	59	113	163	201	219	225	225	224	160
May	48	97	150	189	211	220	215	207	204	192
June	62	114	162	193	213	216	208	196	193	203
July	49	96	148	186	207	215	211	202	200	192
Aug.	27	58	109	157	193	211	217	217	217	159
Sept.	19	19	43	103	148	189	213	224	227	101
Oct.	11	11	11	40	83	135	167	191	199	46
Nov.	4	4	4	4	15	44	66	87	93	8
Dec.	0	0	0	0	1	5	11	14	15	1

48° N. Lat

	N (Shade)	NNE/NNW	NE/NW	ENE/WNW	E/W	ESE/WSW	SE/SW	SSE/SSW	S	HOR
Jan.	15	15	15	53	118	175	216	239	245	85
Feb.	20	20	36	103	168	216	242	249	250	138
Mar.	26	26	80	154	204	234	239	232	228	188
Apr.	31	61	132	180	219	225	215	194	186	226
May	35	97	158	200	218	214	192	163	150	247
June	46	110	165	204	215	206	180	148	134	252
July	37	96	156	196	214	209	187	158	146	244
Aug.	33	61	128	174	211	216	208	188	180	223
Sep.	27	27	72	144	191	223	228	223	220	182
Oct.	21	21	35	96	161	207	233	241	242	136
Nov.	15	15	15	52	115	172	212	234	240	85
Dec.	13	13	13	36	91	156	195	225	233	65

52° N. Lat

	N (Shade)	NNE/NNW	NE/NW	ENE/WNW	E/W	ESE/WSW	SE/SW	SSE/SSW	S	HOR
Jan.	13	13	13	39	92	155	193	222	230	62
Feb.	18	18	29	85	156	202	235	247	250	115
Mar.	24	24	73	145	196	230	239	238	236	169
Apr.	30	56	128	177	215	224	220	204	199	211
May	34	98	154	198	217	217	199	175	167	235
June	45	111	161	202	214	210	188	162	152	242
July	36	97	152	194	213	212	195	171	163	233
Aug.	32	56	124	169	208	216	212	197	193	208
Sep.	25	25	65	136	182	218	228	228	227	163
Oct.	19	19	28	80	148	192	225	238	240	114
Nov.	13	13	13	39	90	152	189	217	225	62
Dec.	10	10	10	19	73	127	172	199	209	42

56° N. Lat

	N (Shade)	NNE/NNW	NE/NW	ENE/WNW	E/W	ESE/WSW	SE/SW	SSE/SSW	S	HOR
Jan.	10	10	10	21	74	126	169	194	205	40
Feb.	16	16	21	71	139	184	223	239	244	91
Mar.	22	22	65	136	185	224	238	241	241	149
Apr.	28	58	123	173	211	223	223	213	210	195
May	36	99	149	195	215	218	206	187	181	222
June	53	111	160	199	213	213	196	174	168	231
July	37	98	147	192	211	214	201	183	177	221
Aug.	30	56	119	165	203	216	215	206	203	193
Sep.	23	23	58	126	171	211	227	230	231	144
Oct.	16	16	20	68	132	176	213	229	234	91
Nov.	10	10	10	21	72	122	165	190	200	40
Dec.	7	7	7	7	47	92	135	159	171	23

Table 7-9. Maximum solar heat gain factor for sunlit glass Btu/sq ft, continued. (Copyright 1989 by the American Society of Heating, Refrigerating and Air-Conditioning Engineers, Inc., from the ASHRAE *Handbook—Fundamentals*. Used by permission.)

Fenestration Facing	Room Construction	Solar Time, h																							
		1	2	3	4	5	6	7	8	9	10	11	12	13	14	15	16	17	18	19	20	21	22	23	24
N (Shaded)	L	0.17	0.14	0.11	0.09	0.08	0.33	0.42	0.48	0.56	0.63	0.71	0.76	0.80	0.82	0.82	0.79	0.75	0.84	0.61	0.48	0.38	0.31	0.25	0.20
	M	0.23	0.20	0.18	0.16	0.14	0.34	0.41	0.46	0.53	0.59	0.65	0.70	0.73	0.75	0.76	0.74	0.75	0.79	0.61	0.50	0.42	0.36	0.31	0.27
	H	0.25	0.23	0.21	0.20	0.19	0.38	0.45	0.49	0.55	0.60	0.65	0.69	0.72	0.72	0.72	0.70	0.70	0.75	0.57	0.46	0.39	0.34	0.31	0.28
NNE	L	0.06	0.05	0.04	0.03	0.03	0.26	0.43	0.47	0.44	0.41	0.40	0.39	0.39	0.38	0.36	0.33	0.30	0.26	0.20	0.16	0.13	0.10	0.08	0.07
	M	0.09	0.08	0.07	0.06	0.06	0.24	0.38	0.42	0.39	0.37	0.37	0.36	0.36	0.36	0.34	0.33	0.30	0.27	0.22	0.18	0.16	0.14	0.12	0.10
	H	0.11	0.10	0.09	0.09	0.08	0.26	0.39	0.42	0.39	0.36	0.35	0.34	0.34	0.33	0.32	0.31	0.28	0.25	0.21	0.18	0.16	0.14	0.13	0.12
NE	L	0.04	0.04	0.03	0.02	0.02	0.23	0.41	0.51	0.51	0.45	0.39	0.36	0.33	0.31	0.28	0.26	0.23	0.19	0.15	0.12	0.10	0.08	0.06	0.05
	M	0.07	0.06	0.06	0.05	0.04	0.21	0.36	0.44	0.45	0.40	0.36	0.33	0.31	0.30	0.28	0.26	0.23	0.21	0.17	0.15	0.13	0.11	0.09	0.08
	H	0.09	0.08	0.08	0.07	0.07	0.23	0.37	0.44	0.44	0.39	0.34	0.31	0.29	0.27	0.26	0.24	0.22	0.20	0.17	0.14	0.13	0.12	0.11	0.10
ENE	L	0.04	0.03	0.03	0.02	0.02	0.21	0.40	0.52	0.57	0.53	0.45	0.39	0.34	0.31	0.28	0.25	0.22	0.18	0.14	0.12	0.09	0.08	0.06	0.05
	M	0.07	0.06	0.05	0.05	0.04	0.20	0.35	0.45	0.49	0.47	0.41	0.36	0.33	0.30	0.28	0.26	0.23	0.20	0.17	0.14	0.12	0.11	0.09	0.08
	H	0.09	0.09	0.08	0.07	0.07	0.22	0.36	0.46	0.49	0.45	0.38	0.33	0.30	0.27	0.25	0.23	0.21	0.19	0.16	0.14	0.13	0.12	0.11	0.10
E	L	0.04	0.03	0.03	0.02	0.02	0.19	0.37	0.51	0.57	0.57	0.50	0.42	0.37	0.32	0.29	0.25	0.22	0.19	0.15	0.12	0.10	0.08	0.06	0.05
	M	0.07	0.06	0.06	0.05	0.05	0.18	0.33	0.44	0.50	0.51	0.46	0.39	0.35	0.31	0.29	0.26	0.23	0.21	0.17	0.15	0.13	0.11	0.10	0.08
	H	0.09	0.09	0.08	0.08	0.07	0.20	0.34	0.45	0.49	0.49	0.43	0.36	0.32	0.29	0.26	0.24	0.22	0.19	0.17	0.15	0.13	0.12	0.11	0.10
ESE	L	0.05	0.04	0.03	0.03	0.02	0.17	0.34	0.49	0.58	0.61	0.57	0.48	0.41	0.36	0.32	0.28	0.24	0.20	0.16	0.13	0.10	0.09	0.07	0.06
	M	0.08	0.07	0.06	0.05	0.05	0.16	0.31	0.43	0.51	0.54	0.51	0.44	0.39	0.35	0.32	0.29	0.26	0.22	0.19	0.16	0.14	0.12	0.11	0.09
	H	0.10	0.09	0.09	0.08	0.08	0.19	0.32	0.43	0.50	0.52	0.49	0.41	0.36	0.32	0.29	0.26	0.24	0.21	0.18	0.16	0.14	0.13	0.12	0.11
SE	L	0.05	0.04	0.04	0.03	0.03	0.13	0.28	0.43	0.55	0.62	0.63	0.57	0.48	0.42	0.37	0.33	0.28	0.24	0.19	0.15	0.12	0.10	0.08	0.07
	M	0.09	0.08	0.07	0.06	0.05	0.14	0.26	0.38	0.48	0.54	0.56	0.51	0.45	0.40	0.36	0.33	0.29	0.25	0.21	0.18	0.16	0.14	0.12	0.10
	H	0.11	0.10	0.10	0.09	0.08	0.17	0.28	0.40	0.49	0.53	0.53	0.48	0.41	0.36	0.33	0.30	0.27	0.24	0.20	0.18	0.16	0.14	0.13	0.12
SSE	L	0.07	0.05	0.04	0.04	0.03	0.06	0.15	0.29	0.43	0.55	0.63	0.64	0.60	0.52	0.45	0.40	0.35	0.29	0.23	0.18	0.15	0.12	0.10	0.08
	M	0.11	0.09	0.08	0.07	0.06	0.08	0.16	0.26	0.38	0.48	0.55	0.57	0.54	0.48	0.43	0.39	0.35	0.30	0.25	0.21	0.18	0.16	0.14	0.12
	H	0.12	0.11	0.11	0.10	0.09	0.12	0.19	0.29	0.40	0.49	0.54	0.55	0.51	0.44	0.39	0.35	0.31	0.27	0.23	0.20	0.18	0.16	0.15	0.13
S	L	0.08	0.07	0.05	0.04	0.04	0.06	0.09	0.14	0.22	0.34	0.48	0.59	0.65	0.65	0.59	0.50	0.43	0.36	0.28	0.22	0.18	0.15	0.12	0.10
	M	0.12	0.11	0.09	0.08	0.07	0.08	0.11	0.14	0.21	0.31	0.42	0.52	0.57	0.58	0.53	0.47	0.41	0.36	0.29	0.25	0.21	0.18	0.16	0.14
	H	0.13	0.12	0.12	0.11	0.10	0.11	0.14	0.17	0.24	0.33	0.43	0.51	0.56	0.55	0.50	0.43	0.37	0.32	0.26	0.22	0.20	0.18	0.16	0.15
SSW	L	0.10	0.08	0.07	0.06	0.05	0.06	0.09	0.11	0.15	0.19	0.27	0.39	0.52	0.62	0.67	0.65	0.58	0.46	0.36	0.28	0.23	0.19	0.15	0.12
	M	0.14	0.12	0.11	0.09	0.08	0.09	0.11	0.13	0.15	0.18	0.25	0.35	0.46	0.55	0.59	0.59	0.53	0.44	0.35	0.30	0.25	0.22	0.19	0.16
	H	0.15	0.14	0.13	0.12	0.11	0.12	0.14	0.16	0.18	0.21	0.27	0.37	0.46	0.53	0.57	0.55	0.49	0.40	0.32	0.26	0.23	0.20	0.18	0.16
SW	L	0.12	0.10	0.08	0.07	0.05	0.06	0.08	0.10	0.12	0.14	0.16	0.24	0.36	0.49	0.60	0.66	0.66	0.58	0.43	0.33	0.27	0.22	0.18	0.14
	M	0.15	0.14	0.12	0.10	0.09	0.09	0.10	0.12	0.13	0.15	0.17	0.23	0.33	0.44	0.53	0.58	0.59	0.53	0.41	0.33	0.28	0.24	0.21	0.18
	H	0.15	0.14	0.13	0.12	0.11	0.12	0.13	0.14	0.16	0.17	0.19	0.25	0.34	0.44	0.52	0.56	0.56	0.49	0.37	0.30	0.25	0.21	0.19	0.17
WSW	L	0.12	0.10	0.08	0.07	0.05	0.06	0.07	0.09	0.10	0.12	0.13	0.17	0.26	0.40	0.52	0.62	0.66	0.61	0.44	0.34	0.27	0.22	0.18	0.15
	M	0.15	0.13	0.12	0.10	0.09	0.09	0.10	0.11	0.12	0.13	0.14	0.17	0.24	0.35	0.46	0.54	0.58	0.55	0.42	0.34	0.28	0.24	0.21	0.18
	H	0.15	0.14	0.13	0.12	0.11	0.11	0.12	0.13	0.14	0.15	0.16	0.19	0.26	0.36	0.46	0.53	0.56	0.51	0.38	0.30	0.25	0.21	0.19	0.17
W	L	0.12	0.10	0.08	0.06	0.05	0.06	0.07	0.08	0.10	0.11	0.12	0.14	0.20	0.32	0.45	0.57	0.64	0.61	0.44	0.34	0.27	0.22	0.18	0.14
	M	0.15	0.13	0.11	0.10	0.09	0.09	0.09	0.10	0.11	0.12	0.13	0.14	0.19	0.29	0.40	0.50	0.56	0.55	0.41	0.33	0.27	0.23	0.20	0.17
	H	0.14	0.13	0.12	0.11	0.10	0.11	0.12	0.13	0.14	0.14	0.15	0.16	0.21	0.30	0.40	0.49	0.54	0.52	0.38	0.30	0.24	0.21	0.20	0.17
WNW	L	0.12	0.10	0.08	0.06	0.05	0.06	0.07	0.09	0.10	0.12	0.13	0.15	0.17	0.26	0.40	0.53	0.63	0.62	0.44	0.34	0.27	0.22	0.18	0.14
	M	0.15	0.13	0.11	0.10	0.09	0.09	0.10	0.11	0.12	0.13	0.14	0.15	0.17	0.24	0.35	0.47	0.55	0.55	0.41	0.33	0.27	0.23	0.20	0.17
	H	0.14	0.13	0.12	0.11	0.10	0.11	0.12	0.13	0.14	0.15	0.16	0.17	0.18	0.25	0.36	0.46	0.53	0.52	0.38	0.30	0.24	0.20	0.18	0.16
NW	L	0.11	0.09	0.08	0.06	0.05	0.06	0.08	0.10	0.12	0.14	0.16	0.17	0.19	0.23	0.33	0.47	0.59	0.60	0.42	0.33	0.26	0.21	0.17	0.14
	M	0.14	0.12	0.11	0.09	0.08	0.09	0.10	0.11	0.13	0.14	0.16	0.17	0.18	0.21	0.30	0.42	0.51	0.54	0.39	0.32	0.26	0.22	0.19	0.16
	H	0.14	0.12	0.11	0.10	0.10	0.10	0.12	0.13	0.15	0.16	0.18	0.18	0.19	0.22	0.30	0.41	0.50	0.51	0.36	0.29	0.23	0.20	0.17	0.15
NNW	L	0.12	0.09	0.08	0.06	0.05	0.07	0.11	0.14	0.18	0.22	0.25	0.27	0.29	0.30	0.33	0.44	0.57	0.62	0.44	0.33	0.26	0.21	0.17	0.14
	M	0.15	0.13	0.11	0.10	0.09	0.10	0.12	0.15	0.18	0.21	0.23	0.26	0.27	0.28	0.31	0.39	0.51	0.56	0.41	0.33	0.27	0.23	0.20	0.17
	H	0.14	0.13	0.12	0.11	0.10	0.12	0.15	0.17	0.20	0.23	0.25	0.26	0.28	0.28	0.31	0.38	0.49	0.53	0.38	0.30	0.25	0.21	0.18	0.16
HOR	L	0.11	0.09	0.07	0.06	0.05	0.07	0.14	0.24	0.36	0.48	0.58	0.66	0.72	0.74	0.73	0.67	0.59	0.47	0.37	0.29	0.24	0.19	0.16	0.13
	M	0.16	0.14	0.12	0.11	0.09	0.11	0.16	0.24	0.33	0.43	0.52	0.59	0.64	0.67	0.66	0.62	0.56	0.47	0.38	0.32	0.28	0.24	0.21	0.18
	H	0.17	0.16	0.15	0.14	0.13	0.15	0.20	0.28	0.36	0.45	0.52	0.59	0.62	0.64	0.62	0.58	0.51	0.42	0.35	0.29	0.26	0.23	0.21	0.19

L = Light construction: frame exterior wall, 2-in. concrete floor slab, approximately 30 lb of material/ft^2 of floor area.
M = Medium construction: 4-in. concrete exterior wall, 4-in. concrete floor slab, approximately 70 lb of building material/ft^2 of floor area.
H = Heavy construction: 6-in. concrete exterior wall, 6-in. concrete floor slab, approximately 130 lb of building materials/ft^2 of floor area.

Table 7-10. Cooling load factors for glass without interior shading. ((Copyright 1989 by the American Society of Heating, Refrigerating and Air-Conditioning Engineers, Inc., from the ASHRAE Handbook—Fundamentals. Used by permission.)

For rooms that have carpeted floors, CLF values should be obtained from Table 7-11. Since the carpeting tends to shade and insulate the floor mass, the heat is converted to a cooling load much more quickly and with a higher peak than a bare or tiled floor. The carpeting prevents the floor from absorbing much of the radiant heat.

Similarly, for spaces that have interior shading on the fenestration areas, CLF values should be obtained from Table 7-12. The CLF values in

	Room												Solar Time												
Dir.	Mass	0100	0200	0300	0400	0500	0600	0700	0800	0900	1000	1100	1200	1300	1400	1500	1600	1700	1800	1900	2000	2100	2200	2300	2400
	L	.00	.00	.00	.00	.01	.64	.73	.74	.81	.88	.95	.98	.98	.94	.88	.79	.79	.55	.31	.12	.04	.02	.01	.00
N	M	.03	.02	.02	.02	.02	.64	.69	.69	.77	.84	.91	.94	.95	.91	.86	.79	.79	.56	.32	.16	.10	.07	.05	.04
	H	.10	.09	.08	.07	.07	.62	.64	.64	.71	.77	.83	.87	.88	.85	.81	.75	.76	.55	.34	.22	.17	.15	.13	.11
	L	.00	.00	.00	.00	.01	.51	.83	.88	.72	.47	.33	.27	.24	.23	.20	.18	.14	.09	.03	.01	.00	.00	.00	.00
NE	M	.01	.01	.00	.00	.01	.50	.78	.82	.67	.44	.32	.28	.26	.24	.22	.19	.15	.11	.05	.03	.02	.02	.01	.01
	H	.03	.03	.03	.02	.03	.47	.71	.72	.59	.40	.30	.27	.26	.25	.23	.20	.17	.13	.08	.06	.05	.05	.04	.04
	L	.00	.00	.00	.00	.00	.42	.76	.91	.90	.75	.51	.30	.22	.18	.16	.13	.11	.07	.02	.01	.00	.00	.00	.00
E	M	.01	.01	.00	.00	.01	.41	.72	.86	.84	.71	.48	.30	.24	.21	.18	.16	.13	.09	.04	.03	.02	.01	.01	.01
	H	.03	.03	.03	.02	.02	.39	.66	.76	.74	.63	.43	.29	.24	.22	.20	.18	.15	.12	.08	.06	.05	.05	.04	04
	L	.00	.00	.00	.00	.00	.27	.58	.81	.93	.93	.81	.59	.37	.27	.21	.18	.14	.09	.03	.01	.00	.00	.00	.00
SE	M	.01	.01	.01	.00	.01	.26	.55	.77	.88	.87	.76	.56	.37	.29	.24	.20	.16	.11	.05	.04	.03	.02	.02	.01
	H	.04	.04	.03	.03	.03	.26	.51	.69	.78	.78	.68	.51	.35	.29	.25	.22	.19	.15	.09	.08	.07	.06	.05	.05
	L	.00	.00	.00	.00	.00	.07	.15	.23	.39	.62	.82	.94	.93	.88	.76	.59	.38	.26	.16	.06	.02	.01	.00	.00
S	M	.01	.01	.01	.00	.01	.07	.14	.22	.38	.59	.78	.88	.88	.82	.76	.57	.38	.28	.18	.09	.06	.04	.03	.02
	H	.05	.05	.04	.04	.03	.09	.15	.21	.35	.54	.70	.79	.79	.69	.52	.37	.29	.21	.13	.10	.09	.08	.07	.06
	L	.00	.00	.00	.00	.00	.04	.09	.13	.16	.19	.23	.39	.62	.82	.94	.94	.81	.54	.19	.07	.03	.01	.00	.00
SW	M	.02	.02	.01	.01	.01	.05	.09	.13	.16	.19	.22	.38	.60	.78	.89	.89	.77	.52	.20	.10	.07	.05	.04	.03
	H	.07	.06	.05	.05	.04	.07	.11	.14	.16	.18	.21	.35	.55	.71	.80	.79	.69	.48	.20	.14	.11	.10	.08	.07
	L	.00	.00	.00	.00	.00	.03	.07	.10	.13	.15	.16	.18	.31	.55	.78	.92	.93	.73	.25	.10	.04	.01	.01	.00
W	M	.02	.02	.01	.01	.01	.04	.07	.10	.13	.14	.16	.17	.30	.53	.74	.87	.88	.69	.24	.12	.07	.05	.04	.03
	H	.06	.06	.05	.04	.04	.06	.09	.11	.13	.15	.16	.17	.28	.49	.67	.78	.79	.62	.23	.14	.11	.09	.08	.07
	L	.00	.00	.00	.00	.00	.04	.09	.14	.17	.20	.22	.23	.24	.31	.53	.78	.92	.81	.28	.10	.04	.02	.01	.00
NW	M	.02	.02	.01	.01	.01	.05	.10	.13	.17	.19	.21	.22	.22	.30	.52	.75	.88	.77	.26	.12	.07	.05	.04	.03
	H	.06	.05	.05	.04	.04	.07	.11	.14	.17	.19	.20	.21	.22	.28	.48	.68	.79	.69	.23	.14	.10	.09	.08	.07
	L	.00	.00	.00	.00	.00	.08	.25	.45	.64	.80	.91	.97	.97	.91	.80	.64	.44	.23	.08	.03	.01	.00	.00	.00
Hor.	M	.02	.02	.01	.01	.01	.08	.24	.43	.60	.75	.86	.92	.92	.87	.77	.63	.45	.26	.12	.07	.05	.04	.03	.02
	H	.07	.06	.05	.05	.04	.11	.25	.41	.56	.68	.77	.83	.83	.80	.71	.59	.44	.28	.17	.13	.11	.10	.09	.08

Values for nominal 15 ft by 15 ft by 10 ft. high space, with ceiling, and 50% or less glass in exposed surface at listed orientation.

L = Lightweight construction, such as 1-in. wood floor, Group G wall.
M = Mediumweight construction, such as 2 to 4-in. concrete floor, Group E wall.
H = Heavyweight construction, such as 6 to 8-in. concrete floor, Group C wall.

Table 7-11. CLFs for glass without shading (carpeted floors). (Copyright 1989 by the American Society of Heating, Refrigerating and Air-Conditioning Engineers, Inc., from the ASHRAE *Handbook—Fundamentals.* Used by permission.)

Table 7-12 account for the fact that part of the incoming radiant energy is reflected back out through the glass and part of the radiant energy is absorbed by the interior shade itself. Most of the radiant energy absorbed by the interior shade is converted to an instantaneous cooling load on the space.

Example 7-4.

Assume that a 4-ft by 4-ft west-facing window, with two layers of clear glazing and 1/4-in. air space between the layers of glazing, is in the exterior wall of a building located in St. Louis, as shown in Figure 7-6. Assume that there is a 2-ft wide overhang located 2 ft above the top of the window; it is solar noon on June 15. Assume also that there is a 7-1/2 mph wind outside, and the inside temperature is 78°F. The building floor is an uncarpeted, 6-in. thick, 120-lb/cu ft concrete floor.

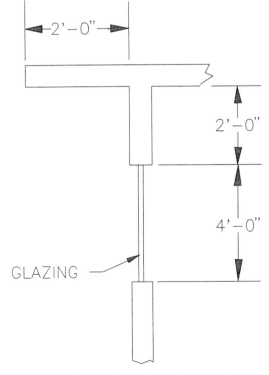

Figure 7-6. Window and overhang for Example 7-4.

Fenes-tration Facing	0100	0200	0300	0400	0500	0600	0700	0800	0900	1000	1100	1200	1300	1400	1500	1600	1700	1800	1900	2000	2100	2200	2300	2400
N	0.08	0.07	0.06	0.06	0.07	0.73	0.66	0.65	0.73	0.80	0.86	0.89	0.89	0.86	0.82	0.75	0.78	0.91	0.24	0.18	0.15	0.13	0.11	0.10
NNE	0.03	0.03	0.02	0.02	0.03	0.64	0.77	0.62	0.42	0.37	0.37	0.37	0.36	0.35	0.32	0.28	0.23	0.17	0.08	0.07	0.06	0.05	0.04	0.04
NE	0.03	0.02	0.02	0.02	0.02	0.56	0.76	0.74	0.58	0.37	0.29	0.27	0.26	0.24	0.22	0.20	0.16	0.12	0.06	0.05	0.04	0.04	0.03	0.03
ENE	0.03	0.02	0.02	0.02	0.02	0.52	0.76	0.80	0.71	0.52	0.31	0.26	0.24	0.22	0.20	0.18	0.15	0.11	0.06	0.05	0.04	0.04	0.03	0.03
E	0.03	0.02	0.02	0.02	0.02	0.47	0.72	0.80	0.76	0.62	0.41	0.27	0.24	0.22	0.20	0.17	0.14	0.11	0.06	0.05	0.05	0.04	0.03	0.03
ESE	0.03	0.03	0.02	0.02	0.02	0.41	0.67	0.79	0.80	0.72	0.54	0.34	0.27	0.24	0.21	0.19	0.15	0.12	0.07	0.06	0.05	0.04	0.04	0.03
SE	0.03	0.03	0.02	0.02	0.02	0.30	0.57	0.74	0.81	0.79	0.68	0.49	0.33	0.28	0.25	0.22	0.18	0.13	0.08	0.07	0.06	0.05	0.04	0.04
SSE	0.04	0.03	0.03	0.03	0.02	0.12	0.31	0.54	0.72	0.81	0.81	0.71	0.54	0.38	0.32	0.27	0.22	0.16	0.09	0.08	0.07	0.06	0.05	0.04
S	0.04	0.04	0.03	0.03	0.03	0.09	0.16	0.23	0.38	0.58	0.75	0.83	0.80	0.68	0.50	0.35	0.27	0.19	0.11	0.09	0.08	0.07	0.06	0.05
SSW	0.05	0.04	0.04	0.03	0.03	0.09	0.14	0.18	0.22	0.27	0.43	0.63	0.78	0.84	0.80	0.66	0.46	0.25	0.13	0.11	0.09	0.08	0.07	0.06
SW	0.05	0.05	0.04	0.04	0.03	0.07	0.11	0.14	0.16	0.19	0.22	0.38	0.59	0.75	0.83	0.81	0.69	0.45	0.16	0.12	0.10	0.09	0.07	0.06
WSW	0.05	0.05	0.04	0.04	0.03	0.07	0.10	0.12	0.14	0.16	0.17	0.23	0.44	0.64	0.78	0.84	0.78	0.55	0.16	0.12	0.10	0.09	0.07	0.06
W	0.05	0.05	0.04	0.04	0.03	0.06	0.09	0.11	0.13	0.15	0.16	0.17	0.31	0.53	0.72	0.82	0.81	0.61	0.16	0.12	0.10	0.08	0.07	0.06
WNW	0.05	0.05	0.04	0.03	0.03	0.07	0.10	0.12	0.14	0.16	0.17	0.18	0.22	0.43	0.65	0.80	0.84	0.66	0.16	0.12	0.10	0.08	0.07	0.06
NW	0.05	0.04	0.04	0.03	0.03	0.07	0.11	0.14	0.17	0.19	0.20	0.21	0.22	0.30	0.52	0.73	0.82	0.69	0.16	0.12	0.10	0.08	0.07	0.06
NNW	0.05	0.05	0.04	0.03	0.03	0.11	0.17	0.22	0.26	0.30	0.32	0.33	0.34	0.34	0.39	0.61	0.82	0.76	0.17	0.12	0.10	0.08	0.07	0.06
HOR.	0.06	0.05	0.04	0.04	0.03	0.12	0.27	0.44	0.59	0.72	0.81	0.85	0.85	0.81	0.71	0.58	0.42	0.25	0.14	0.12	0.10	0.08	0.07	0.06

Table 7-12. CLFs for glass with interior shading. (Copyright 1989 by the American Society of Heating, Refrigerating and Air-Conditioning Engineers, Inc., from the ASHRAE Handbook—Fundamentals. Used by permission.)

What is the cooling load due to the window? What would the cooling load be if carpeting were installed? What would the cooling load be if venetian blinds were installed?

Apply Equation 7.4 to calculate the solar angle of declination, or use the list of solar declination values on page 71:

$$\delta = 23.1 \text{ degrees for the month of June}$$

A latitude of approximately 38 degrees is obtained from Table 3-1 for St. Louis, MO.

Use Equation 7.5 to calculate the solar altitude for 38-degrees N latitude:

$$\beta = \sin^{-1}(\sin(23.1) \sin(38) + \cos(23.1) \cos(38) \cos(0)) = 68.4 \text{ degrees}$$

Use Equation 7.7 and Figure 7-6 to calculate the shading factor or fraction:

$$S = \left(\frac{1}{4 \text{ ft}}\right)(2 \text{ ft} \tan(68.4) - 2 \text{ ft}) = 0.76$$

Window shaded area = (4 ft x 4 ft)(0.76) = 12.2 sq ft

Unshaded area = (4 ft x 4 ft)(1 - 0.76) = 3.8 sq ft

Calculate the shading coefficient for clear, double glazed glass with a 7-1/2 mph wind (h = 4.0) from Table 7-6:

$$SC = 0.81$$

From Table 4-2 for double glazed glass:

$$U = 0.61 \text{ Btuh/sq ft-}°F$$

Find the CLTD value for glass at 12 noon solar time from Table 7-8:

$$CLTD = 9°F$$

Using Equation 7.1, calculate the conduction heat gain:

$$q_c = (0.61 \text{ Btuh/sq ft-}°F)(16 \text{ sq ft})(9°F) = 87.8 \text{ Btuh}$$

Table 7-9 lists solar heat gain factors for north latitudes in 4-degree increments. Since St. Louis, MO is at 38-degrees N latitude and the nearest tabulated values are for 36-degrees and 40-degrees, the value for either latitude may be used. In this case, 40-degrees N is chosen.

For a 6-in. thick, 120-lb/cu ft concrete floor, assume heavy construction. For the direct solar radiation, unshaded glass at 40-degrees N latitude, west-facing in June, from Table 7-9, the maximum solar heat gain is:

$$SHGF = 216 \text{ Btuh/sq ft-}°F$$

From Table 7-10, west fenestration, heavy construction, 12 noon solar time:

$$CLF = 0.16$$

Use Equation 7.9 to calculate the radiation heat transfer as follows:

$$q_r = (3.8 \text{ sq ft})(0.81)(216 \text{ Btuh/sq ft-}°F)(0.16)$$
$$= 106.4 \text{ Btuh}$$

To calculate the indirect radiation for the shaded part of the window, use the values for north-facing fenestration:

$$\text{Maximum SHGF} = 48 \text{ Btuh/sq ft-}°F$$

$$CLF = 0.69$$

Use Equation 7.9 to calculate the cooling load for the direct radiation:

$$q_r = (12.2 \text{ sq ft})(0.81)(48 \text{ Btuh/sq ft-}°F)(0.69)$$
$$= 327.3 \text{ Btuh}$$

Therefore, the total radiant and convective cooling load for the window is:

$$q = 87.8 \text{ Btuh} + 106.4 \text{ Btuh} + 327.3 \text{ Btuh}$$
$$= 521.5 \text{ Btuh}$$

For a room with carpeting, the conduction cooling load through the glass would be the same:

$$q_c = 87.8 \text{ Btuh}$$

For the unshaded west glass, at solar noon, from Table 7-11:

$$CLF = 0.17$$

Using Equation 7.9 to calculate the radiant cooling load again:

$$q_r = (3.8 \text{ sq ft})(0.81 \text{ Btuh/sq ft-}°F)(216)(0.17)$$
$$= 113.0 \text{ Btuh}$$

For the shaded west glass, at solar noon, from Table 7-10:

$$CLF = 0.87$$

$$q_r = (12.2 \text{ sq ft})(0.81)(48 \text{ Btuh/sq ft-}°F)(0.87)$$
$$= 412.7 \text{ Btuh}$$

Therefore, the cooling load with carpeting is:

$$q = 87.8 \text{ Btuh} + 113.0 \text{ Btuh} + 412.7 \text{ Btuh}$$
$$= 613.5 \text{ Btuh}$$

For a fenestration area with venetian blinds (indoor shade), from Table 4-2, U = 0.55 Btuh/sq ft-°F. Using Equation 7.1, the conduction cooling load is:

$$q_c = (0.55 \text{ Btuh/sq ft-°F})(16 \text{ sq ft})(9°F) = 79.2 \text{ Btuh}$$

For the unshaded west glass at solar noon (from Table 7-12):

$$CLF = 0.17$$

$$q_r = (3.8 \text{ sq ft})(0.81)(216 \text{ Btuh/sq ft-°F})(0.17)$$
$$= 113.0 \text{ Btuh}$$

For the shaded area using the values for north (from Table 7-12):

$$CLF = 0.89$$

$$q_r = (12.2 \text{ sq ft})(0.81)(48 \text{ Btuh/sq ft-°F})(0.89)$$
$$= 422.2 \text{ Btuh}$$

The total radiant and convective cooling load for the shaded window is:

$$q = 79.2 \text{ Btuh} + 113.0 \text{ Btuh} + 422.2 \text{ Btuh}$$
$$= 614.4 \text{ Btuh}$$

Cooling loads from unconditioned spaces

Cooling loads can also result from heat transmission through surfaces adjacent to unconditioned spaces. Unconditioned spaces can include areas like garages, mechanical or electrical rooms, store rooms, boiler rooms, and kitchens that are not directly cooled.

Cooling loads for unconditioned spaces are essentially steady-state loads and, therefore, may be calculated in a manner similar to heat losses. Equation 4.1 may be used to calculate heat transmission through surfaces adjacent to unconditioned spaces. Table 7-13 lists typical recommended temperature differences between conditioned and unconditioned spaces.

For spaces other than those listed in Table 7-13, for which a temperature difference is not known, a temperature difference of 5°F may be assumed.

Exterior cooling load calculation procedure

Most spaces have multiple heat sources that all make up the total peak cooling load. Each has a peak cooling load that frequently occurs at

Item	Temperature difference* (°F)
Partitions	10
Partitions, or glass in partitions, adjacent to laundries, kitchens, or boiler rooms	25
Floors above unconditioned rooms	10
Floors on ground	0
Floors above basements	0
Floors above rooms or basements used as laundries, kitchens, or boiler rooms	35
Floors above vented spaces	17
Floors above unvented spaces	0
Ceilings with unconditioned rooms above	10
Ceilings with rooms above used as laundries, kitchens, etc.	20

* The temperature differences are based on the assumption that the air conditioning system is being designed to maintain an inside temperature 17°F lower than the outdoor temperature. For air conditioning systems designed to maintain a greater temperature difference than 17°F between the inside and outside, add to the values in the above table, the difference between the assumed design temperature difference and 17°F.

Table 7-13. Design temperature differences. (Extracted From Table 3.1, Reference 6.)

different times. For this reason, it is necessary to perform cooling load calculations for a number of different times or hours. Since calculations are frequently necessary for several different times, in order to determine the actual peak, an External Cooling Load Calculation form has been provided to help organize the calculations.

When selecting the times for which cooling loads are to be calculated, it is usually necessary to use some judgment in order to determine which cooling loads are the dominant loads for the space. One method that may be used to determine the dominant loads is to calculate the U x A (overall heat transfer coefficient times the surface area) value for each exterior surface. The surface (or surfaces) with the largest U x A value frequently are the dominant loads.

Another method is to look at exterior surfaces with large areas and light construction. Sometimes those are the dominant loads.

A relatively large fenestration area will frequently dominate cooling loads for a space. (Note that the maximum solar heat gain for fenestration areas can occur in the winter months.) Finally, interior cooling loads, which will be discussed in Chapter 8, should be examined to determine if they are the largest cooling load for the space.

Once the dominant space cooling load or loads are identified, the times at which these loads peak are usually a good basis for selecting the times for which calculations are made. Once a probable peak time is selected, it should be entered in the center column of the three time columns on the External Cooling Load Calculation form. The columns to the left and right are assigned times that are 1 or 2 hours different than the probable peak time, with the left column being earlier and the right column later. These, then, are the hours for which calculations should be made in order to estimate the actual peak space cooling load.

The procedure for estimating the peak external cooling load and the hour at which it occurs is as follows:

1. Select the interior design dry bulb and wet bulb temperatures or the dry bulb temperature and relative humidity.

2. Select the outdoor design conditions from Table 3-1 for the appropriate building location. Select the design month or months.

3. Calculate the net surface area of each exterior surface such as walls, roofs, fenestration areas, etc.

4. Calculate the U value and weight per square foot of all opaque exterior surfaces. Calculate a U value for surfaces adjacent to unconditioned spaces. Select an appropriate U value from Table 4-2 for each type of fenestration.

5. Select the appropriate wall group from Table 7-1, and/or roof-ceiling group from Table 7-5, for each exterior wall and roof assembly.

6. Estimate the time at which the peak space cooling load will occur, and select the times for which calculations should be made based on the above suggestions.

7. Obtain the CLTD values at the selected calculation hours from Table 7-2 for sunlit walls and/or Table 7-5 for sunlit roofs.

8. Obtain the latitude-month correction factor (LM) from Table 7-3; determine a color correction factor (if any). Determine the appropriate indoor, outdoor, and average outdoor temperature correction factors (if any). Determine the ventilation correction factor (if any). Use Equation 7.2 for walls and/or Equation 7.3 for roofs. Calculate the corrected CLTD values at each calculation hour for each exterior wall or roof.

9. Using Equation 7.1, calculate the sensible cooling load for each opaque surface for each hour.

10. From Table 7-8, obtain CLTD values for the calculation times and multiply them by the fenestration area to estimate the conduction cooling loads for glass.

11. Calculate the net area of all surfaces adjacent to unconditioned spaces. Using the recommended temperature differences in Table 7-13, calculate the cooling loads

for each surface adjacent to an unconditioned space.

12. Obtain SCs for each type of fenestration area from Table 7-7.

13. Using the methods described in the section on shading and Equation 7.7, or in References 1 and 4, calculate the shaded and unshaded areas for fenestration area.

14. Obtain a maximum solar heat gain factor (SHGF) for each fenestration orientation from Table 7-9 for the appropriate month and latitude. Use the values for north-facing glass for all shaded areas of fenestration.

15. Obtain a CLF for each calculation time for each fenestration orientation from Table 7-10, 7-11, or 7-12 for the appropriate design situation.

16. Using Equation 7.9, calculate the radiant cooling load for each fenestration area for each calculation time.

17. Total the individual cooling loads to get the exterior cooling load subtotal. These subtotals are then combined with the interior cooling loads to get the total peak cooling loads.

This procedure is outlined in the table on the following page.

EXTERNAL COOLING LOADS

CONDUCTION

ITEM	TYPE	ORIENT	U	AREA	U X A	CLTD TABLE VALUE			ADJUST FOR LATITUDE & MONTH	COLOR CORRECT K	ADJUST FOR INSIDE & OUTSIDE TEMPS	CORR. CLTD			SENSIBLE COOLING LOAD		
						_HR	_HR	_HR				__HR	__HR	__HR	___HR	___HR	___HR
ROOF																	
ROOF																	
ROOF																	
WALL																	
WALL																	
WALL																	
WALL																	
WALL																	
DOOR																	
DOOR																	
DOOR																	

ITEM	TYPE	U	AREA	U X A	TEMP. DIFF.	U X A X T
PARTITION						
PARTITION						
DOOR						
DOOR						
GLASS						
CEILING						
FLOOR						
SUM						

SOLAR		ORIENT	TOTAL AREA	SHADE COEFF.	MAX. SHGF	__HR		__HR		__HR	
						AREA	CLF	AREA	CLF	AREA	CLF
GLASS	UNSHADED										
	SHADED										
GLASS	UNSHADED										
	SHADED										
GLASS	UNSHADED										
	SHADED										
GLASS	UNSHADED										
	SHADED										

SUBTOTAL

Chapter 8

Internal Heat Gains and Cooling Loads

Internal heat gains are generated within the conditioned space itself. Typical examples include gains from people, lights, and energy-consuming equipment located within the space.

Cooling loads from transient heat transfer and the effect of thermal mass was discussed in Chapters 4 and 7. The reader was also introduced to cooling load temperature differences (CLTDs) and cooling load factors (CLFs).

ASHRAE developed CLFs for internal heat gains to account for the thermal storage of the building mass. Buildings have been classified into light-mass, medium-mass, and heavy-mass construction groups. Cooling loads and CLFs for interior heat gains are primarily a function of the length of time since the heat gain began, or entered, the space and the thermal mass of the space.

As sensible heat is absorbed by the building mass, the temperature of the mass increases. As the temperature of the mass increases, it is able to absorb less heat, causing the cooling load to approach instantaneous heat gain. After the heat gain stops, or leaves the space, the thermal mass continues to reject heat to the space, causing a cooling load.

When calculating cooling loads that result from internal heat gains, a diversity factor is occasionally applied to the loads. Diversity of cooling loads results from not all of the cooling load being on or used at any particular time. Some

rooms may be occupied while others are not. For large spaces or buildings, some of the lights may be on while some are off. Diversity factors account for variable loads and can range from 70% to 100% depending on the size and use of the building. The application of diversity factors requires some knowledge of the operation of a building and should be applied with considerable judgment. For small spaces, it is likely that all internal gains will be on; therefore, a diversity factor of 100% should be used when calculating cooling loads for small spaces.

Cooling loads from people

People (occupants) are one of the few sources of indoor heat that generate both sensible and latent heat. Heat is generated within a person's body by an oxidation process known as metabolism. Heat generated by humans is a function of the meta-bolic rate, which is the rate at which food is converted to energy in the form of heat. The metabolic rate is a function of the person's activity level. The more active a person is, the more heat they generate and transfer to the space. The more sedentary a person is, the less they generate heat.

Sensible heat from occupants is transferred to a space as a result of the difference in temperature between the surrounding air and the body temperature. To a lesser extent, sensible heat also is transferred by radiant heat exchange with the surroundings. Latent heat is transferred from people in the form of perspiration. The perspiration is then evaporated to the air in the space.

Since all of the latent heat is absorbed directly by the air as moisture, it is an instantaneous heat gain or cooling load for the space. Unlike sensible heat, there is no storage of latent heat by the building mass and furnishings. Therefore, the instantaneous latent heat gain is also a latent cooling load, and no cooling factors need be applied. Instantaneous sensible and latent heat gains for occupants are given in Table 8-1.

Sensible heat gains, and the resultant cooling loads on a space, are a function of the number of occupants in the space, their activity level, the type of clothing worn, and how long the occupants have been in the space. The length of time they have been in the space has an effect on the

cooling loads, since it also affects the thermal storage of the space.

The values given in Table 8-1 are for an average male, weighing 150 lb. Assume a room air temperature of 75°F and an occupancy of at least 3 hours. These values may be used without the need for correction unless there are significant differences between the actual and assumed conditions listed above. For a room dry bulb air temperature of 80°F, the total heat rejection would remain the same, but the sensible gain should be decreased approximately 20% while the latent values are increased accordingly.

The sex, weight, and age of a person will also affect the body heat rejection rate for any given activity level. Typically, the metabolic rate for females is about 85% of the values in Table 8-1, and about 75% for children. An example of such

DEGREE OF ACTIVITY	TYPICAL APPLICATION	Metabolic Rate (Adult Male) Btu/hr	Average Adjusted Metabolic Rate* Btu/hr	ROOM DRY-BULB TEMPERATURE									
				82 F BTU/hr		80 F BTU/hr		78 F BTU/hr		75 F BTU/hr		70 F BTU/hr	
				Sensible	Latent	Sensible	Latent	Sensible	Latent	Sensible	Latent	Sensible	Latent
Seated at rest	Theatre, Grade School	390	350	175	175	195	155	210	140	230	120	260	90
Seated, very light work	High School	450	400	180	220	195	205	215	185	240	160	275	125
Office worker	Offices, Hotels, Apts., College	475	450	180	270	200	250	215	235	245	205	285	165
Standing, walking slowly	Dept., Retail or Variety Store	550											
Walking, seated	Drug Store	550	500	180	320	200	300	220	280	255	245	290	210
Standing, walking slowly	Bank	550											
Sedentary work	Restaurant	500	550	190	360	220	330	240	310	280	270	320	230
Light bench work	Factory, light work	800	750	190	560	200	530	245	505	295	455	365	385
Moderate dancing	Dance Hall	900	850	220	630	245	605	275	575	325	525	400	450
Walking, 3 mph	Factory, fairly heavy work	1000	1000	270	730	300	700	330	670	380	620	460	540
Heavy Work	Bowling Alley, Factory	1500	1450	450	1000	465	985	485	965	525	925	605	845

Adjusted Metabolic Rate is the metabolic rate to be applied to a mixed group of people with a typical percent composition based on the following factors:
Metabolic rate, adult female=Metabolic rate, adult male x 0.85
Metabolic rate, children =Metabolic rate, adult male x 0.75
Restaurant - Values for this application include 60 Btu per hr for food per individual (30 Btu sensible and 30 Btu latent heat per hour).
Bowling - Assume one person per alley actually bowling and all others sitting, metabolic rate 400 Btu per hr; or standing, 550 Btu per hr.

Table 8-1. Heat gain from people. (Courtesy, Carrier Corporation, Copyright 1965, McGraw-Hill, Inc. Used with permission.)

a situation might occur when the majority of the occupants are small children. It would be up to the designers to decide the extent of the adjustment, if any.

The tabulated values for people eating in restaurants were increased to account for the sensible and latent loads caused by the food served.

Other animals also give off heat, which is directly proportional to their weight, for any given activity. For sedentary activity, the total heat rejected per animal (sensible and latent) may be estimated as follows:

$$q = 6.6 \ W^{0.75}$$

Equation 8.1

where:

q = total heat rejected (Btuh)
W = weight of the animal (lb)

In order to account for the thermal storage effect of the space, a CLF must be applied to the instantaneous sensible heat gain for occupants from Table 8-1. Equation 8.2 may be used to calculate the sensible cooling load for occupants:

$$q_o = (q_i)(n)(f)(CLF)$$

Equation 8.2

where:

q_o = sensible cooling load for occupants (Btuh)
q_i = instantaneous sensible heat gain for each occupant from Table 8-1 or Equation 8.1 (Btuh/person)
n = number of occupants
f = adjustment for age, sex, or weight
CLF = cooling load factor from Table 8-2

Notice that the CLF values in Table 8-2 are a function of how long the occupants will be in the space, and how long it has been since they entered the space.

Example 8-1.

Assume there are 10 boys playing basketball in a gymnasium, with two adult female coaches and one adult male referee. They have been in the gym for 2 hours and will be there for a total of 4 hours. The space temperature is 75°F. What are the sensible, latent, and total cooling loads?

- Assume heavy work level of activity for the children with a 0.75 age adjustment factor.

- Assume a standing or walking activity level and a 0.85 sex adjustment factor for the female coaches.

- Assume a 3-mph walking activity level for the male referee. No age or sex adjustment is required.

Total hours in space	Hours After Each Entry Into Space																							
	1	2	3	4	5	6	7	8	9	10	11	12	13	14	15	16	17	18	19	20	21	22	23	24
2	0.49	0.58	0.17	0.13	0.10	0.08	0.07	0.06	0.05	0.04	0.04	0.03	0.03	0.02	0.02	0.02	0.02	0.01	0.01	0.01	0.01	0.01	0.01	0.01
4	0.49	0.59	0.66	0.71	0.27	0.21	0.16	0.14	0.11	0.10	0.08	0.07	0.06	0.06	0.05	0.04	0.04	0.03	0.03	0.03	0.02	0.02	0.02	0.01
6	0.50	0.60	0.67	0.72	0.76	0.79	0.34	0.26	0.21	0.18	0.15	0.13	0.11	0.10	0.08	0.07	0.06	0.06	0.05	0.04	0.04	0.03	0.03	0.03
8	0.51	0.61	0.67	0.72	0.76	0.80	0.82	0.84	0.38	0.30	0.25	0.21	0.18	0.15	0.13	0.12	0.10	0.09	0.08	0.07	0.06	0.05	0.05	0.04
10	0.53	0.62	0.69	0.74	0.77	0.80	0.83	0.85	0.87	0.89	0.42	0.34	0.28	0.23	0.20	0.17	0.15	0.13	0.11	0.10	0.09	0.08	0.07	0.06
12	0.55	0.64	0.70	0.75	0.79	0.81	0.84	0.86	0.88	0.89	0.91	0.92	0.45	0.36	0.30	0.25	0.21	0.19	0.16	0.14	0.12	0.11	0.09	0.08
14	0.58	0.66	0.72	0.77	0.80	0.83	0.85	0.87	0.89	0.90	0.91	0.92	0.93	0.94	0.47	0.38	0.31	0.26	0.23	0.20	0.17	0.15	0.13	0.11
16	0.62	0.70	0.75	0.79	0.82	0.85	0.87	0.88	0.90	0.91	0.92	0.93	0.94	0.95	0.95	0.96	0.49	0.39	0.33	0.28	0.24	0.20	0.18	0.16
18	0.66	0.74	0.79	0.82	0.85	0.87	0.89	0.90	0.92	0.93	0.94	0.94	0.95	0.96	0.96	0.97	0.97	0.97	0.50	0.40	0.33	0.28	0.24	0.21

CLF = 1.0 for systems shut down at night and for high occupant densities such as in theaters and auditoriums.

Table 8-2. Sensible CLFs for people. (Copyright 1989 by the American Society of Heating, Refrigerating and Air-Conditioning Engineers, Inc., from the ASHRAE *Handbook—Fundamentals.* Used by permission.)

From Table 8-1 and using Equation 8.2, the cooling loads are as follows:

Occupants	Sensible heat gain	
Boys:		
(525 Btuh/boy)(10 boys)		
(0.75 adj. factor)	=	3,938 Btuh
Coaches:		
(245 Btuh/coach)(2 coaches)		
(0.85 adj. factor)	=	416 Btuh
Referee:		
(380 Btuh/referee)		
(1 referee)(1)	=	380 Btuh
Sensible subtotal	=	**4,734 Btuh**

Occupants	Latent heat gain	
Boys:		
(925 Btuh/boy)(10 boys)		
(0.75 adj. factor)	=	6,938 Btuh
Coaches:		
(205 Btuh/coach)(2 coaches)		
(0.85 adj. factor)	=	349 Btuh
Referee:		
(620 Btuh/referee)		
(1 referee)(1)	=	620 Btuh
Latent subtotal	=	**7,907 Btuh**

From Table 8-2, the sensible CLF = 0.59

Therefore, the cooling loads are as follows:

Sensible: (4,734)(0.59)	=	2,793	Btuh
Latent:	=	7,907	Btuh
Total:	=	10,700	Btuh

Cooling loads from lights

Cooling loads from lighting fixtures represent one of the largest loads, either external or internal, for many types of buildings such as offices, retail spaces, and any space with a large number of light fixtures. In office buildings, lighting levels can be 2 to 3 watts per sq ft of floor area. All of the electrical energy supplied to a light fixture is eventually converted to heat.

The heat given off is either radiant (visible light), heat convected to the surrounding air, or heat conducted to adjacent building materials.

In a manner similar to heat gains for sunlit walls and roofs, the visible radiant energy (light) is absorbed by the space itself and the furnishings within the space. The rate at which electricity is converted to heat gain by a light fixture is a function of a number of variables including the lamp type (incandescent, fluorescent, etc.), the type of fixture (recessed, pendant mounted, etc.), and the efficiency of the light fixture itself.

Lighting fixtures can also be vented or unvented. A vented fixture is one that has air slots in it allowing supply air or return air to pass through the slots. A typical recessed unvented fixture is shown in Figure 8-1, while a typical recessed vented fixture is shown in Figure 8-2.

Incandescent lights convert about 10% of the power input to visible light; about 90% is lost as radiant, conductive, and convective heat. Fluorescent lights convert about 25% of the power input to visible light, with the balance of the energy supplied lost as heat. Part of the energy loss for a fluorescent fixture is by the ballast, which amounts to about 20% of the energy supplied to the fixture.

All heat rejected by a light fixture is sensible, there is no latent heat. The instantaneous heat gain from lights may be calculated with the following equation:

$$q_s = (P_l)(D)(BF)(3.413)$$

Equation 8.3

where:

q_s = instantaneous heat gain (Btuh)

P_l = lighting power input (watts)

D = a diversity or use factor to account for not all lights being on at once.

BF = ballast factor; BF = 1.0 for incandescent light, and BF = 1.2 for fluorescent light

3.413 = conversion from watts to Btuh

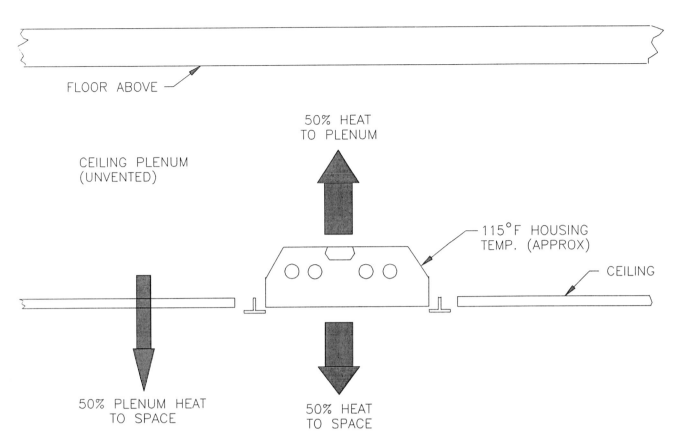

Figure 8-1. Typical recessed unvented luminaire installation.

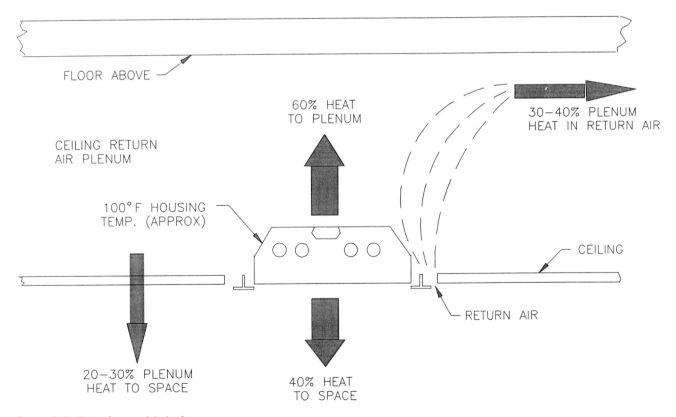

Figure 8-2. Typical vented light fixture.

In Equation 8.3, note that the power is expressed in lighting power input. Sometimes, a lighting fixture is described by the total power input, including losses. An example of this situation is a 200-watt, four-tube fluorescent fixture. The fixture has four 40-watt tubes for a total of 160 watts. The other 40 watts are for ballast losses. If the 200-watt value was used for the fixture wattage in Equation 8.3, the BF would equal 1, since the ballast losses were already accounted for in the fixture rating.

The relative amount of heat that is either stored in the building mass and furnishings or converted to an instantaneous cooling load, depends upon the type of fixture mounting, space ventilation rate, and building thermal mass.

ASHRAE has developed CLFs for lights which are similar to those for people. To determine a lighting CLF, the fixture type must be known, an approximate space ventilation rate must be known or estimated, the space construction weight must be estimated, the number of hours the lights have been on must be determined, and the total length of time the lights will be on must be known.

Table 8-3 lists an "a" classification or coefficient for lights. This classification is determined by the features of the room furnishing the light fixture type and the approximate space ventilation rate. Table 8-4 lists a "b" classification or value for lights. This classification is determined by the envelope or building construction and the space ventilation rates.

After an "a" or "b" classification has been determined from Tables 8-3 and 8-4, a CLF for the lights may then be determined from Table 8-5. Note that Table 8-5 is comprised of multiple tables based on the length of time the lights are on.

Example 8-2.

Assume a room has four recessed 200-watt fluorescent light fixtures (non-ventilated type). The lights were turned on at 8:00 a.m. and will be on until 6:00 p.m. The room has carpeting over a 6-in. concrete floor with a desk, chair, and bookcase. It is expected that the room will have a medium ventilation rate. What is the cooling load from the lights at 2:00 p.m.?

"a" Classification	Furnishings	Air Supply and Return	Type of Light Fixture
0.45	Heavyweight, simple furnishings, no carpet	Low rate, supply and return below ceiling	Recessed, not vented
0.55	Ordinary furniture, no carpet	Medium to high ventilation rate, supply and return below ceiling or through ceiling grille and space	Recessed, not vented
0.65	Ordinary furniture, with or without carpet	Medium to high ventilation rate or fan coil or induction type air conditioning terminal unit; supply through ceiling or wall diffuser; return around light fixtures and through ceiling space	Vented
0.75	Any type of furniture	Ducted returns through light fixtures	Vented or free hanging in air stream with ducted returns

Table 8-3. Design values of "a" coefficient, features of room furnishings, light fixtures, and ventilation rates.

Room envelope construction	Mass of floor area (lb/sq ft)	Room air circulation and type of supply and return			
		Low	Medium	High	Very high
2-in. wood floor	10	B	A	A	A
3-in. concrete floor	40	B	B	B	A
6-in. concrete floor	75	C	C	C	B
8-in. concrete floor	120	D	D	C	C
12-in. concrete floor	160	D	D	D	D

Table 8-4. Calculated for different envelope construction and room air ventilation rates.

It is assumed that the wattage specified is the nominal power input per fixture. Therefore, the 200 watts include a ballast factor. For a single room, the diversity factor is assumed to be 1.0. From Equation 8.3, the instantaneous heat gain is as follows:

$$q_i = (200 \text{ W})(1.0)(1.0)(3.413) = 683 \text{ Btuh}$$

Referring to Table 8-3, an "a" value of 0.55 is selected as the closest match, even though it is for no carpeting. It does match the ventilation rate and fixture type.

Referring to Table 8-4, a "b" value of c is chosen for a 6-in. concrete floor with a medium ventilation rate.

The lights were turned on at 8:00 a.m. and will be on until 6:00 p.m., which is 10 hours total. At 2:00 p.m. the lights have been on for 6 hours. Referring to Table 8-5, CLF = 0.75. Therefore, the cooling load is:

$$q_s = (CLF)(q_i) = (0.75)(683) = 512 \text{ Btuh}$$

Cooling loads for equipment and appliances

Internal heat gains and the resulting cooling loads for equipment, come from heat-producing equipment located within the conditioned space. Equipment heat gains come from such items as office equipment (typewriters, copy machines,

computers, etc.); appliances (ranges, stoves, ovens, etc.); and from motor-driven equipment (pumps, compressors, machines, etc.).

All of the above equipment produces sensible heat only, except the kitchen appliances, which can also produce latent heat. In an effort to reduce the heat gain to a space, exhaust hoods are placed over kitchen appliances to remove heat before it enters the space. For cooling load calculation purposes, appliances are classified according to whether or not they are located under an exhaust hood (hooded or unhooded). These classifications apply to all equipment, but most frequently apply to kitchen cooking appliances.

In a manner similar to people and lights, ASHRAE has developed CLFs for equipment and appliances located within the conditioned space. Table 8-6 lists CLFs for hooded appliances. Table 8-7 lists factors for unhooded appliances and equipment.

As in the case of CLFs for lights, CLF values for equipment depend upon the total time the appliances are operated and how long they have been on. The CLFs should be applied to the sensible portion of the heat gain only. They do not apply to latent heat gain. The following equation may be used to calculate sensible cooling loads for equipment and appliances:

$$q = (q_i)(D)(CLF)$$

Equation 8.4

Cooling Load Factors When Lights Are on for 8 Hours

"a" Coefficients	"b" Classification	0	1	2	3	4	5	6	7	8	9	10	11	12	13	14	15	16	17	18	19	20	21	22	23
												Number of hours after lights are turned on													
0.45	A	0.02	0.46	0.57	0.65	0.72	0.77	0.82	0.85	0.88	0.46	0.37	0.30	0.24	0.19	0.15	0.12	0.10	0.08	0.06	0.05	0.04	0.03	0.03	0.02
	B	0.07	0.51	0.56	0.61	0.65	0.68	0.71	0.74	0.77	0.34	0.31	0.28	0.25	0.22	0.20	0.18	0.16	0.15	0.13	0.12	0.11	0.10	0.09	0.08
	C	0.11	0.55	0.58	0.60	0.63	0.65	0.67	0.69	0.71	0.28	0.26	0.25	0.23	0.22	0.20	0.19	0.18	0.17	0.16	0.15	0.14	0.13	0.12	0.12
	D	0.14	0.58	0.60	0.61	0.62	0.63	0.64	0.65	0.66	0.22	0.22	0.21	0.20	0.20	0.19	0.19	0.18	0.18	0.17	0.16	0.16	0.16	0.15	0.15
0.55	A	0.01	0.56	0.65	0.72	0.77	0.82	0.85	0.88	0.90	0.37	0.30	0.24	0.19	0.16	0.13	0.10	0.08	0.07	0.05	0.04	0.03	0.03	0.02	0.02
	B	0.06	0.60	0.64	0.68	0.71	0.74	0.76	0.79	0.81	0.28	0.25	0.23	0.20	0.18	0.16	0.15	0.13	0.12	0.11	0.10	0.09	0.08	0.07	0.06
	C	0.09	0.63	0.66	0.68	0.70	0.71	0.73	0.75	0.76	0.23	0.21	0.20	0.19	0.18	0.17	0.16	0.15	0.14	0.13	0.12	0.11	0.11	0.10	0.10
	D	0.11	0.66	0.67	0.68	0.69	0.70	0.71	0.72	0.72	0.18	0.18	0.17	0.17	0.16	0.16	0.15	0.15	0.14	0.14	0.13	0.13	0.13	0.12	0.12
0.65	A	0.01	0.66	0.73	0.78	0.82	0.86	0.88	0.91	0.93	0.29	0.23	0.19	0.15	0.12	0.10	0.08	0.06	0.05	0.04	0.03	0.03	0.02	0.02	0.01
	B	0.04	0.69	0.72	0.75	0.77	0.80	0.82	0.84	0.85	0.22	0.19	0.18	0.16	0.14	0.13	0.12	0.10	0.09	0.08	0.08	0.07	0.06	0.06	0.05
	C	0.07	0.72	0.73	0.75	0.76	0.78	0.79	0.80	0.82	0.18	0.17	0.16	0.15	0.14	0.13	0.12	0.11	0.11	0.10	0.10	0.09	0.08	0.08	0.07
	D	0.09	0.73	0.74	0.75	0.76	0.77	0.77	0.78	0.79	0.14	0.14	0.13	0.13	0.13	0.12	0.12	0.11	0.11	0.11	0.10	0.10	0.10	0.10	0.09
0.75	A	0.01	0.76	0.80	0.84	0.87	0.90	0.92	0.93	0.95	0.21	0.17	0.13	0.11	0.09	0.07	0.06	0.05	0.04	0.03	0.02	0.02	0.02	0.01	0.01
	B	0.03	0.78	0.80	0.82	0.84	0.85	0.87	0.88	0.89	0.15	0.14	0.13	0.11	0.10	0.09	0.08	0.07	0.07	0.06	0.05	0.05	0.04	0.04	0.04
	C	0.05	0.80	0.81	0.82	0.83	0.84	0.85	0.86	0.87	0.13	0.12	0.11	0.10	0.10	0.09	0.09	0.08	0.08	0.07	0.07	0.06	0.06	0.06	0.05
	D	0.06	0.81	0.82	0.82	0.83	0.83	0.84	0.84	0.85	0.10	0.10	0.10	0.09	0.09	0.09	0.08	0.08	0.08	0.08	0.07	0.07	0.07	0.07	0.07

Cooling Load Factors When Lights Are on for 10 Hours

"a" Coefficients	"b" Classification	0	1	2	3	4	5	6	7	8	9	10	11	12	13	14	15	16	17	18	19	20	21	22	23
												Number of hours after lights are turned on													
0.45	A	0.03	0.47	0.58	0.66	0.73	0.78	0.82	0.86	0.88	0.91	0.93	0.49	0.39	0.32	0.26	0.21	0.17	0.13	0.11	0.09	0.07	0.06	0.05	0.04
	B	0.10	0.54	0.59	0.63	0.66	0.70	0.73	0.76	0.78	0.80	0.82	0.39	0.35	0.32	0.28	0.26	0.23	0.21	0.19	0.17	0.15	0.14	0.12	0.11
	C	0.15	0.59	0.61	0.64	0.66	0.68	0.70	0.72	0.73	0.75	0.76	0.33	0.31	0.29	0.27	0.26	0.24	0.23	0.21	0.20	0.19	0.18	0.17	0.16
	D	0.18	0.62	0.63	0.64	0.66	0.67	0.68	0.69	0.69	0.70	0.71	0.27	0.26	0.26	0.25	0.24	0.23	0.23	0.22	0.21	0.21	0.20	0.19	0.19
0.55	A	0.02	0.57	0.65	0.72	0.78	0.82	0.85	0.88	0.91	0.92	0.94	0.40	0.32	0.26	0.21	0.17	0.14	0.11	0.09	0.07	0.06	0.05	0.04	0.03
	B	0.08	0.62	0.66	0.69	0.73	0.75	0.78	0.80	0.82	0.84	0.85	0.32	0.29	0.26	0.23	0.21	0.19	0.17	0.15	0.14	0.12	0.11	0.10	0.09
	C	0.12	0.66	0.68	0.70	0.72	0.74	0.75	0.77	0.78	0.79	0.81	0.27	0.25	0.24	0.22	0.21	0.20	0.19	0.17	0.16	0.15	0.14	0.14	0.13
	D	0.15	0.69	0.70	0.71	0.72	0.73	0.73	0.74	0.75	0.76	0.76	0.22	0.22	0.21	0.20	0.20	0.19	0.18	0.18	0.17	0.17	0.16	0.16	0.15
0.65	A	0.02	0.66	0.73	0.78	0.83	0.86	0.89	0.91	0.93	0.94	0.95	0.31	0.25	0.20	0.16	0.13	0.11	0.08	0.07	0.05	0.04	0.04	0.03	0.02
	B	0.06	0.71	0.74	0.76	0.79	0.81	0.83	0.84	0.86	0.87	0.89	0.25	0.22	0.20	0.18	0.16	0.15	0.13	0.12	0.11	0.10	0.09	0.08	0.07
	C	0.09	0.74	0.75	0.77	0.78	0.80	0.81	0.82	0.83	0.84	0.85	0.21	0.20	0.18	0.17	0.16	0.15	0.14	0.14	0.13	0.12	0.11	0.11	0.10
	D	0.11	0.76	0.77	0.77	0.78	0.79	0.79	0.80	0.81	0.81	0.82	0.17	0.17	0.16	0.16	0.15	0.15	0.14	0.14	0.14	0.13	0.13	0.12	0.12
0.75	A	0.01	0.76	0.81	0.84	0.88	0.90	0.92	0.93	0.95	0.96	0.97	0.22	0.18	0.14	0.12	0.09	0.08	0.06	0.05	0.04	0.03	0.03	0.02	0.02
	B	0.04	0.79	0.81	0.83	0.85	0.86	0.88	0.89	0.90	0.91	0.92	0.18	0.16	0.14	0.13	0.12	0.10	0.09	0.08	0.08	0.07	0.06	0.06	0.05
	C	0.07	0.81	0.82	0.83	0.84	0.85	0.86	0.87	0.88	0.89	0.89	0.15	0.14	0.13	0.12	0.12	0.11	0.10	0.10	0.09	0.09	0.08	0.08	0.07
	D	0.08	0.83	0.83	0.84	0.84	0.85	0.85	0.86	0.86	0.87	0.87	0.12	0.12	0.12	0.11	0.11	0.11	0.10	0.10	0.10	0.10	0.09	0.09	0.09

Cooling Load Factors When Lights Are on for 12 Hours

"a" Coefficients	"b" Classification	0	1	2	3	4	5	6	7	8	9	10	11	12	13	14	15	16	17	18	19	20	21	22	23
												Number of hours after lights are turned on													
0.45	A	0.05	0.49	0.59	0.67	0.73	0.78	0.83	0.86	0.89	0.91	0.93	0.94	0.95	0.51	0.41	0.33	0.27	0.22	0.17	0.14	0.11	0.09	0.07	0.06
	B	0.13	0.57	0.61	0.65	0.69	0.72	0.75	0.77	0.79	0.82	0.83	0.85	0.87	0.43	0.39	0.35	0.31	0.28	0.25	0.23	0.21	0.18	0.17	0.15
	C	0.19	0.63	0.65	0.67	0.69	0.71	0.73	0.74	0.76	0.77	0.79	0.80	0.81	0.37	0.35	0.33	0.31	0.29	0.27	0.26	0.24	0.23	0.21	0.20
	D	0.22	0.66	0.67	0.68	0.69	0.70	0.71	0.72	0.73	0.74	0.74	0.75	0.76	0.32	0.31	0.30	0.29	0.28	0.27	0.26	0.26	0.25	0.24	0.23
0.55	A	0.04	0.58	0.66	0.73	0.78	0.82	0.86	0.89	0.91	0.93	0.94	0.95	0.96	0.42	0.34	0.27	0.22	0.18	0.14	0.11	0.09	0.07	0.06	0.05
	B	0.11	0.65	0.68	0.72	0.74	0.77	0.79	0.81	0.83	0.85	0.86	0.88	0.89	0.35	0.32	0.28	0.26	0.23	0.21	0.19	0.17	0.15	0.14	0.12
	C	0.15	0.69	0.71	0.73	0.75	0.76	0.78	0.79	0.80	0.81	0.83	0.84	0.85	0.30	0.29	0.27	0.25	0.24	0.22	0.21	0.20	0.19	0.17	0.16
	D	0.18	0.72	0.73	0.74	0.75	0.76	0.76	0.77	0.78	0.78	0.79	0.80	0.80	0.26	0.25	0.24	0.24	0.23	0.22	0.22	0.21	0.20	0.20	0.19
0.65	A	0.03	0.67	0.74	0.79	0.83	0.86	0.89	0.91	0.93	0.94	0.95	0.96	0.97	0.33	0.26	0.21	0.17	0.14	0.11	0.09	0.07	0.06	0.05	0.04
	B	0.09	0.73	0.75	0.78	0.80	0.82	0.84	0.85	0.87	0.88	0.89	0.90	0.91	0.27	0.25	0.22	0.20	0.18	0.16	0.15	0.13	0.12	0.11	0.10
	C	0.12	0.76	0.78	0.79	0.80	0.81	0.83	0.84	0.85	0.86	0.86	0.87	0.88	0.24	0.22	0.21	0.20	0.19	0.17	0.16	0.15	0.14	0.14	0.13
	D	0.14	0.79	0.79	0.80	0.80	0.81	0.82	0.82	0.83	0.83	0.84	0.84	0.85	0.20	0.20	0.19	0.18	0.18	0.17	0.17	0.16	0.16	0.15	0.15
0.75	A	0.02	0.77	0.81	0.85	0.88	0.90	0.92	0.94	0.95	0.96	0.97	0.97	0.98	0.23	0.19	0.15	0.12	0.10	0.08	0.06	0.05	0.04	0.03	0.03
	B	0.06	0.81	0.82	0.84	0.86	0.87	0.88	0.90	0.91	0.92	0.92	0.93	0.94	0.19	0.18	0.16	0.14	0.13	0.12	0.10	0.09	0.08	0.08	0.07
	C	0.09	0.83	0.84	0.85	0.86	0.87	0.88	0.88	0.89	0.90	0.90	0.91	0.91	0.17	0.16	0.15	0.14	0.13	0.12	0.12	0.11	0.10	0.10	0.09
	D	0.10	0.85	0.85	0.86	0.86	0.86	0.87	0.87	0.88	0.88	0.88	0.89	0.89	0.14	0.14	0.14	0.13	0.13	0.12	0.12	0.12	0.11	0.11	0.11

Table 8-5. Cooling load factors for lights. (Copyright 1989 by the American Society of Heating, Refrigerating and Air-Conditioning Engineers, Inc., from the ASHRAE Handbook—Fundamentals. Used by permission.)

where:

q = sensible cooling load (Btuh)

q_i = instantaneous sensible heat gain (Btuh)

D = a diversity or use factor

CLF = CLF from Table 8-6 or 8-7

Whenever possible, the manufacturer of the equipment should be contacted to get the actual instantaneous heat gain or heat rejection rate. If that is not possible, the tables in the following sections of this chapter may be used. The data in these tables are average values for typical equipment. It may also be possible to check the equipment nameplate for heat rejection rates or power consumption rates.

Cooling Load Factors When Lights Are on for 14 Hours

"a"Coef-ficients	"b"Class-ification	0	1	2	3	4	5	6	7	8	9	10	11	12	13	14	15	16	17	18	19	20	21	22	23
	A	0.07	0.51	0.61	0.68	0.74	0.79	0.83	0.87	0.89	0.91	0.93	0.94	0.95	0.96	0.97	0.53	0.42	0.34	0.27	0.22	0.18	0.14	0.12	0.09
	B	0.18	0.61	0.65	0.68	0.72	0.74	0.77	0.79	0.81	0.83	0.85	0.86	0.88	0.89	0.90	0.46	0.41	0.37	0.34	0.30	0.27	0.24	0.22	0.20
0.45	C	0.24	0.67	0.69	0.71	0.73	0.74	0.76	0.77	0.79	0.80	0.81	0.82	0.83	0.84	0.85	0.41	0.39	0.36	0.34	0.32	0.30	0.28	0.27	0.25
	D	0.26	0.71	0.72	0.72	0.73	0.74	0.75	0.76	0.77	0.78	0.78	0.79	0.80	0.80	0.80	0.36	0.35	0.34	0.33	0.32	0.31	0.30	0.29	0.28
	A	0.06	0.69	0.68	0.74	0.79	0.83	0.86	0.89	0.91	0.93	0.94	0.95	0.96	0.97	0.98	0.43	0.35	0.28	0.22	0.18	0.15	0.12	0.10	0.08
	B	0.15	0.68	0.71	0.74	0.77	0.79	0.81	0.83	0.85	0.86	0.88	0.89	0.90	0.91	0.92	0.38	0.34	0.31	0.27	0.25	0.22	0.20	0.18	0.16
0.55	C	0.19	0.73	0.75	0.76	0.78	0.79	0.80	0.81	0.83	0.84	0.85	0.86	0.86	0.87	0.88	0.34	0.32	0.30	0.28	0.26	0.25	0.23	0.22	0.21
	D	0.22	0.76	0.77	0.77	0.78	0.79	0.79	0.80	0.81	0.81	0.82	0.82	0.83	0.83	0.84	0.29	0.28	0.28	0.27	0.26	0.25	0.24	0.24	0.23
	A	0.05	0.69	0.75	0.80	0.84	0.87	0.89	0.92	0.93	0.95	0.96	0.96	0.97	0.98	0.98	0.34	0.27	0.22	0.17	0.14	0.11	0.09	0.07	0.06
	B	0.11	0.75	0.78	0.80	0.82	0.64	0.85	0.87	0.88	0.89	0.90	0.91	0.92	0.93	0.94	0.29	0.26	0.24	0.21	0.19	0.17	0.16	0.14	0.13
0.65	C	0.15	0.79	0.80	0.82	0.83	0.84	0.85	0.86	0.86	0.87	0.88	0.89	0.89	0.90	0.91	0.26	0.25	0.23	0.22	0.20	0.19	0.18	0.17	0.16
	D	0.17	0.81	0.82	0.82	0.83	0.83	0.84	0.84	0.85	0.85	0.86	0.86	0.87	0.87	0.87	0.23	0.22	0.21	0.21	0.20	0.20	0.19	0.18	0.18
	A	0.03	0.78	0.82	0.86	0.88	0.91	0.92	0.94	0.95	0.96	0.97	0.97	0.98	0.98	0.99	0.24	0.19	0.16	0.12	0.10	0.08	0.07	0.05	0.04
	B	0.08	0.82	0.84	0.86	0.87	0.88	0.90	0.91	0.92	0.92	0.93	0.94	0.94	0.95	0.96	0.21	0.19	0.17	0.15	0.14	0.12	0.11	0.10	0.09
0.75	C	0.11	0.85	0.86	0.87	0.88	0.88	0.89	0.90	0.90	0.91	0.91	0.92	0.92	0.93	0.93	0.19	0.18	0.17	0.16	0.15	0.14	0.13	0.12	0.11
	D	0.12	0.87	0.87	0.87	0.88	0.88	0.89	0.89	0.89	0.90	0.90	0.90	0.90	0.90	0.91	0.16	0.16	0.15	0.15	0.14	0.14	0.14	0.13	0.13

Cooling Load Factors When Lights Are on for 16 Hours

"a"Coef-ficients	"b"Class-ification	0	1	2	3	4	5	6	7	8	9	10	11	12	13	14	15	16	17	18	19	20	21	22	23
	A	0.12	0.54	0.63	0.70	0.76	0.81	0.85	0.88	0.90	0.92	0.94	0.95	0.96	0.97	0.97	0.98	0.98	0.54	0.43	0.35	0.28	0.23	0.18	0.15
	B	0.23	0.66	0.69	0.72	0.75	0.78	0.80	0.82	0.84	0.85	0.87	0.88	0.89	0.90	0.91	0.92	0.93	0.49	0.44	0.39	0.35	0.32	0.29	0.26
0.45	C	0.29	0.72	0.74	0.75	0.77	0.78	0.80	0.81	0.82	0.83	0.84	0.85	0.86	0.87	0.88	0.88	0.89	0.45	0.42	0.39	0.37	0.35	0.33	0.31
	D	0.31	0.75	0.76	0.77	0.77	0.78	0.79	0.79	0.80	0.81	0.81	0.82	0.82	0.83	0.83	0.84	0.84	0.40	0.39	0.37	0.36	0.35	0.34	0.33
	A	0.10	0.63	0.70	0.76	0.81	0.84	0.87	0.90	0.92	0.93	0.95	0.96	0.97	0.97	0.98	0.98	0.99	0.44	0.35	0.28	0.23	0.18	0.15	0.12
	B	0.19	0.72	0.75	0.77	0.80	0.82	0.84	0.85	0.87	0.88	0.89	0.90	0.91	0.92	0.93	0.94	0.94	0.40	0.36	0.32	0.29	0.26	0.24	0.21
0.55	C	0.24	0.77	0.79	0.80	0.81	0.82	0.83	0.84	0.85	0.86	0.87	0.88	0.88	0.89	0.90	0.90	0.91	0.37	0.34	0.32	0.30	0.29	0.27	0.25
	D	0.26	0.80	0.80	0.81	0.82	0.82	0.83	0.83	0.84	0.84	0.85	0.85	0.86	0.86	0.86	0.87	0.87	0.33	0.32	0.31	0.30	0.29	0.28	0.27
	A	0.07	0.71	0.77	0.81	0.85	0.88	0.90	0.92	0.94	0.95	0.96	0.97	0.97	0.98	0.98	0.99	0.99	0.34	0.27	0.22	0.18	0.14	0.12	0.09
	B	0.15	0.78	0.81	0.82	0.84	0.86	0.87	0.88	0.90	0.91	0.92	0.92	0.93	0.94	0.94	0.95	0.96	0.31	0.28	0.25	0.23	0.20	0.18	0.16
0.65	C	0.18	0.82	0.83	0.84	0.85	0.86	0.87	0.88	0.89	0.89	0.90	0.90	0.91	0.92	0.92	0.93	0.93	0.28	0.27	0.25	0.24	0.22	0.21	0.20
	D	0.20	0.84	0.85	0.85	0.86	0.86	0.87	0.87	0.87	0.88	0.88	0.88	0.89	0.89	0.89	0.90	0.90	0.25	0.25	0.24	0.23	0.22	0.22	0.21
	A	0.05	0.79	0.83	0.87	0.89	0.91	0.93	0.94	0.95	0.96	0.97	0.98	0.98	0.98	0.99	0.99	0.99	0.24	0.20	0.16	0.13	0.10	0.08	0.07
	B	0.11	0.85	0.86	0.87	0.89	0.90	0.91	0.92	0.93	0.93	0.94	0.95	0.95	0.96	0.96	0.96	0.96	0.22	0.20	0.18	0.16	0.15	0.13	0.12
0.75	C	0.13	0.87	0.88	0.89	0.89	0.90	0.91	0.91	0.92	0.92	0.93	0.93	0.94	0.94	0.94	0.95	0.95	0.20	0.19	0.18	0.17	0.16	0.15	0.14
	D	0.14	0.89	0.89	0.89	0.90	0.90	0.90	0.91	0.91	0.91	0.91	0.92	0.92	0.92	0.92	0.93	0.93	0.18	0.18	0.17	0.17	0.16	0.16	0.15

CLF = 1.0 when cooling system operates only during occupied hours or when lights are on 24 h/day.

Table 8-5. Cooling load factors for lights, continued. (Copyright 1989 by the American Society of Heating, Refrigerating and Air-Conditioning Engineers, Inc., from the ASHRAE *Handbook—Fundamentals*. Used by permission.)

Sensible Heat Cooling Load Factors for Appliances—Hooded

Total Operational Hours	1	2	3	4	5	6	7	8	9	10	11	12	13	14	15	16	17	18	19	20	21	22	23	24
2	0.27	0.40	0.25	0.18	0.14	0.11	0.09	0.08	0.07	0.06	0.05	0.04	0.04	0.03	0.03	0.03	0.02	0.02	0.02	0.02	0.01	0.01	0.01	0.01
4	0.28	0.41	0.51	0.59	0.39	0.30	0.24	0.19	0.16	0.14	0.12	0.10	0.09	0.08	0.07	0.06	0.05	0.05	0.04	0.04	0.03	0.03	0.02	0.02
6	0.29	0.42	0.52	0.59	0.65	0.70	0.48	0.37	0.30	0.25	0.21	0.18	0.16	0.14	0.12	0.11	0.09	0.08	0.07	0.06	0.05	0.05	0.04	0.04
8	0.31	0.44	0.54	0.61	0.66	0.71	0.75	0.78	0.55	0.43	0.35	0.30	0.25	0.22	0.19	0.16	0.14	0.13	0.11	0.10	0.08	0.07	0.06	0.06
10	0.33	0.46	0.55	0.62	0.68	0.72	0.76	0.79	0.81	0.84	0.60	0.48	0.39	0.33	0.28	0.24	0.21	0.18	0.16	0.14	0.12	0.11	0.09	0.08
12	0.36	0.49	0.58	0.64	0.69	0.74	0.77	0.80	0.82	0.85	0.87	0.88	0.64	0.51	0.42	0.36	0.31	0.26	0.23	0.20	0.18	0.15	0.13	0.12
14	0.40	0.52	0.61	0.67	0.72	0.76	0.79	0.82	0.84	0.86	0.88	0.89	0.91	0.92	0.67	0.54	0.45	0.38	0.32	0.28	0.24	0.21	0.19	0.16
16	0.45	0.57	0.65	0.70	0.75	0.78	0.81	0.84	0.86	0.87	0.89	0.90	0.92	0.93	0.94	0.94	0.69	0.56	0.46	0.39	0.34	0.29	0.25	0.22
18	0.52	0.63	0.70	0.75	0.79	0.82	0.84	0.86	0.88	0.89	0.91	0.92	0.93	0.94	0.95	0.95	0.96	0.96	0.71	0.58	0.48	0.41	0.35	0.30

Table 8-6. CLFs for hooded appliances. (Copyright 1989 by the American Society of Heating, Refrigerating and Air-Conditioning Engineers, Inc., from the ASHRAE *Handbook—Fundamentals*. Used by permission.)

Office equipment

Table 8-8 lists typical heat gains for office equipment. Notice that some of the equipment listed includes a standby value as well as a maximum power input. Equipment such as copy machines are usually not operated at full power continuously but are on standby much of the time they are on. The Recommended Rate of Heat Gain in the last column of Table 8-8 includes a diversity factor to account for standby and full-use time.

Sensible Heat Cooling Load Factors for Appliances—Unhooded

Total Operational Hours	Hours after appliances are on																							
	1	2	3	4	5	6	7	8	9	10	11	12	13	14	15	16	17	18	19	20	21	22	23	24
2	0.56	0.64	0.15	0.11	0.08	0.07	0.06	0.05	0.04	0.04	0.03	0.03	0.02	0.02	0.02	0.02	0.01	0.01	0.01	0.01	0.01	0.01	0.01	0.01
4	0.57	0.65	0.71	0.75	0.23	0.18	0.14	0.12	0.10	0.08	0.07	0.06	0.05	0.05	0.04	0.04	0.03	0.03	0.02	0.02	0.02	0.02	0.01	0.01
6	0.57	0.65	0.71	0.76	0.79	0.82	0.29	0.22	0.18	0.15	0.13	0.11	0.10	0.08	0.07	0.06	0.06	0.05	0.04	0.04	0.03	0.03	0.03	0.02
8	0.58	0.66	0.72	0.76	0.80	0.82	0.85	0.87	0.33	0.26	0.21	0.18	0.15	0.13	0.11	0.10	0.09	0.08	0.07	0.06	0.05	0.04	0.04	0.03
10	0.60	0.68	0.73	0.77	0.81	0.83	0.85	0.87	0.89	0.90	0.36	0.29	0.24	0.20	0.17	0.15	0.13	0.11	0.10	0.08	0.07	0.07	0.06	0.05
12	0.62	0.69	0.75	0.79	0.82	0.84	0.86	0.88	0.89	0.91	0.92	0.93	0.38	0.31	0.25	0.21	0.18	0.16	0.14	0.12	0.11	0.09	0.08	0.07
14	0.64	0.71	0.76	0.80	0.83	0.85	0.87	0.89	0.90	0.92	0.93	0.93	0.94	0.95	0.40	0.32	0.27	0.23	0.19	0.17	0.15	0.13	0.11	0.10
16	0.67	0.74	0.79	0.82	0.85	0.87	0.89	0.90	0.91	0.92	0.93	0.94	0.95	0.96	0.96	0.97	0.42	0.34	0.28	0.24	0.20	0.18	0.15	0.13
18	0.71	0.78	0.82	0.85	0.87	0.89	0.90	0.92	0.93	0.94	0.94	0.95	0.96	0.96	0.97	0.97	0.97	0.98	0.43	0.35	0.29	0.24	0.21	0.18

Table 8-7. CLFs for unhooded appliances. (Copyright 1989 by the American Society of Heating, Refrigerating and Air-Conditioning Engineers, Inc., from the ASHRAE *Handbook—Fundamentals*. Used by permission.)

Appliance	Size	Maximum Input		Standby Input		Recommended Rate of Heat Gain	
		Watts	Btu/h	Watts	Btu/h	Watts	Btu/h
Computer Devices							
Communication/ transmission		1800-4600	6140-15700	1640-2810	5600-9600	1640-2810	5600-9600
Disk drives/mass storage		1000-10000	3400-34100	1000-6600	3400-22400	1000-6600	3400-22400
Microcomputer/ wordprocessor	16-640 kbytes[a]	100-600	340-2050	90-530	300-1800	90-530	300-1800
Minicomputer		2200-6600	7500-15000	2200-6600	7500-15000	2200-6600	7500-15000
Printer (laser)	8 pages/min	870	3000	180	600	300	1000
Printer (Line, high speed	5000-more pages/min	1000-5300	3400-18000	500-2550	2160-9040	730-3800	2500-13000
Tape drives		1200-6500	4100-22200	1000-4700	3500-15000	1000-4700	3500-15000
Terminal		90-200	300-700	80-180	270-600	80-180	270-600
Copiers/Typesetters							
Blue print		1150-12500	3900-42700	500-5000	1700-17000	1150-12500	3900-42700
Copiers (large)	30-67[a] copies/min.	5800-22500	1700-6600	5800-22500	900	3100	1700-6600
Copiers	6-30[a] copies/min.	1570-5800	460-1700	1570-5800	300-900	1000-3100	460-1700
Phototypesetter		1725	5900			1520	5200
Mailprocessing							
Inserting machine 3600-6800 pieces/h		600-3300	2000-11300			390-2150	1300-7300
Labeling machine 1500-30000 pieces/h		600-6600	2000-22500			390-4300	1300-14700
Miscellaneous							
Cash register		60	200			48	160
Cold food/beverage		1150-1920	3900-6600			575-960	1960-3280
Coffee maker	10 cup	1500	5120	sensible		1050	3580
				latent		450	1540
Microwave oven	1 ft[3]	600	2050			400	1360
Paper shredder		250-3000	850-10200			200-2420	680-8250
Water cooler	8 gal/h	700	2400			1750	6000

[a] Input is not proportional to capacity.

Table 8-8. Heat gain from selected office equipment. (Copyright 1989 by the American Society of Heating, Refrigerating and Air-Conditioning Engineers, Inc., from the ASHRAE *Handbook—Fundamentals*. Used by permission.)

Kitchen equipment

Most kitchen appliances produce both sensible and latent heat gains in a conditioned space. The majority of the latent heat gain is from the loss of moisture from foods as they are heated and cooked. In addition to latent heat from cooking foods, gas-burning appliances produce latent heat

as a part of the combustion process. A properly designed and operating kitchen hood exhaust system will remove most of the latent heat from appliances before it enters the conditioned space.

Tables 8-9 and 8-10 list heat gains for unhooded electric, gas-burning, and steam-heated kitchen appliances. If the appliances are located under a

APPLIANCE	OVERALL DIMENSIONS Less Legs and Handles (In.)	TYPE OF CON-TROL	MISCELLANEOUS DATA	MFR MAX RATING BTU/hr	MAIN-TAIN-ING RATE BTU/hr	RECOM HEAT GAIN FOR AVG USE		
						Sensible Heat BTU/hr	Latent Heat BTU/hr	Total Heat BTU/hr
Coffee Brewer - 1/2 gal		Man.		2240	306	900	220	1120
Warmer - 1/2 gal		Man.		306	306	230	90	320
4 Coffee Brewing Units with 4 1/2 gal Tank	20 x 30 x 26H	Auto.	Water heater - 2000 watts Brewers - 2960 watts	16,900		4,800	1,200	6,000
Coffee Urn - 3 gal	15 Dia x 34H	Man.	Black Finish	11900	3000	2600	1700	4300
3 gal	12 x 23 oval x 21H	Auto.	Nickel plated	15300	2600	2200	1500	3700
5 gal	18 Dia x 37H	Auto.	Nickel plated	17000	3600	3400	2300	5700
Doughnut Machine	22 x 22 x 57H	Auto.	Exhaust system to outdoors - 1/2 hp motor	16,000		5,000		5,000
Egg Boiler	10 x 13 x 25H	Man.	Med. ht. - 550 watts Low ht - 275 watts	3,740		1,200	800	2,000
Food Warmer with Plate Warmer per sq ft top surface		Auto.	Insulated, separate heating unit for each pot. Plate warmer in base	1,350	500	350	350	700
Food Warmer without Plate Warmer, per sq ft top surface		Auto.	Ditto, without plate warmer	1,020	400	200	350	550
Fry Kettle - 11 1/2 lb fat	12 Dia x 14H	Auto.		8,840	1,100	1,600	2,400	4,000
Fry Kettle - 25 lb fat	16 x 18 x 12H	Auto.	Frying area 12" x 14"	23,800	2,000	3,800	5,700	9,500
Griddle, Frying	18 x 18 x 8H	Auto.	Frying top 18" x 14"	8,000	2,800	3,100	1,700	4,800
Grille, Meat	14 x 14 x 10H	Auto.	Cooking area 10" x 12"	10,200	1,900	3,900	2,100	6,000
Grille, Sandwich	13 x 14 x 10H	Auto.	Grill area 12" x 12"	5,600	1,900	2,700	700	3,400
Roll Warmer	26 x 17 x 13H	Auto.	One drawer	1,500	400	1,100	100	1,200
Toaster, Continuous	15 x 15 x 28H	Auto.	2 Slices wide - 360 slices/hr	7,500	5,000	5,100	1,300	6,400
Toaster Continuous	20 x 15 x 28H	Auto.	4 Slices wide - 720 slices/hr.	10,200	6,000	6,100	2,600	8,700
Toaster, Pop-Up	6 x 11 x 9H	Auto.	2 Slices	4,150	1,000	2,450	450	2,900
Waffle Iron	12 x 13 x 10H	Auto.	One waffle 7" dia	2,480	600	1,100	750	1,850
Waffle Iron for Ice Cream Sandwich	14 x 13 x 10H	Auto.	12 Cakes, each 2 1/2: x 3 3/4"	7,500	1,500	3,100	2,100	5,200

*If properly designed positive exhaust hood is used, multiply recommended value by .50.

Table 8-9. Heat gain from restaurant appliances, electric. (Courtesy, Carrier Corporation, Copyright 1965, McGraw-Hill, Inc. Used by permission.)

properly designed kitchen hood exhaust system, it may be assumed that all of the latent heat is removed and that 50% of the sensible heat is removed by the exhaust system. Additional appliance heat gains may be found in Reference 1.

Equation 8.4 and the CLFs in Table 8-6 or 8-7 may be used to calculate cooling loads from kitchen appliances.

Motor-driven equipment

Equipment driven by electric motors also produces sensible cooling loads in an air conditioned space. Examples of such equipment include lathes, printing presses, compressors, pumps, etc. Sometimes, the motor and the driven equipment are not both in the conditioned space. In that case, a distinction must be made according to whether only the motor, the machinery, or both are located within the conditioned space. Table 8-11 lists heat gains for motors and motor-driven equipment for all three situations. Equation 8.4 and the CLFs in Table 8-11 may be used to calculate the sensible cooling load.

APPLIANCE	OVERALL DIMENSIONS Less Legs and Handles (In.)	TYPE OF CON-TROL	MISCELLANEOUS DATA	MFR MAX RATING BTU/hr	MAIN-TAIN-ING RATE BTU/hr	RECOM HEAT GAIN FOR AVG USE		
						Sensible Heat BTU/hr	Latent Heat BTU/hr	Total Heat BTU/hr
Coffee Brewer - 1/2 gal Warmer - 1/2 gal		Man. Man.	Combination brewer and warmer	3400 500	500	1350 400	350 100	1700 500
Coffee Brewer Unit with Tank	19 x 30 x 26H		4 Brewers and 4 1/2 gal tank			7,200	1,800	9,000
Coffee Urn - 3 gal	15" Dia x 34H	Auto.	Black Finish	3,200	3,900	2,900	2,900	5,800
Coffee Urn - 3 gal	12 x 23 oval x 21H	Auto.	Nickel Plated		3,400	2,500	2,500	5,000
Coffee Urn - 5 gal	18 Dia. x 37H	Auto.	Nickel Plated		4,700	3,900	3,900	7,800
Food Warmer, Values per sq. ft. top surface		Man.	Water bath type	2,000	900	850	450	1,300
Fry Kettle - 15 lb fat	12 x 20 x 18H	Auto.	Frying area 10 x 10	14,250	3,000	4,200	2,800	7,000
Fry Kettle - 28 lb fat	15 x 35 x 11H	Auto.	Frying area 11 x 16	24,000	4,500	7,200	4,800	12,000
Grill - Broil-O-Grill Top Burner Bottom Burner	22 x 14 x 17H (1.4 sq ft grill surface)	Man.	Insulated 22,000 BTU/hr 15,000 BTU/hr	37,000		14,400	3,600	18,000
Stoves, Short Order - Open Top. Values per sq. ft. top surface		Man.	Ring type burners 12000 to 22000 BTU/ea	14,000		4,200	4,200	8,400
Stoves, Short Order - Closed Top. Values per sq. ft. top surface		Man.	Ring type burners 10000 to 12000 BTU/ea	11,000		3,300	3,300	6,600
Toaster, Continuous	15 x 15 x 28H	Auto.	2 Slices wide - 360 slices/hr	12,000	10,000	7,700	3,300	11,000
STEAM HEATED								
Coffee Urn - 3 gal - 3 gal - 3 gal	15 Dia x 34H 12 x 23 oval x 21H 18 Dia x 37H	Auto. Auto. Auto.	Black finish Nickel plated Nickel plated			2900 2400 3400	1900 1600 2300	4800 4000 5700
Coffee Urn - 3 gal - 3 gal - 3 gal	15 Dia x 34H 12 x 23 oval x 21H 18 Dia x 37H	Man. Man. Man.	Black finish Nickel plated Nickel plated			3100 2600 3700	3100 2600 3700	6200 5200 7400
Food Warmer, per sq ft top surface		Auto.				400	500	900
Food Warmer, per sq ft top surface		Man.				450	1,150	1,500

*If properly designed positive exhaust hood is used, multiply recommended value by .50.

Table 8-10. Heat gain from restaurant appliances, gas burning and steam heating. (Courtesy, Carrier Corporation, Copyright 1965, Mc-Graw-Hill, Inc. Used by permission.)

Infiltration cooling loads

Cooling loads caused by infiltration, or leakage, of outdoor air into a conditioned space are usually treated as internal loads, even though they are from an external source. Methods for estimating the amount of infiltration into or out of a space were discussed in Chapter 5.

Since the infiltration air mixes directly with conditioned air in the space, the heat gain becomes an immediate cooling load on the space. There is no thermal storage; therefore, it is not necessary to apply CLTDs or CLFs.

The total cooling load from infiltration is both sensible and latent heat. The sensible heat gain is a result of the difference in the dry bulb temperatures between the indoor and outdoor air. Equation 6.1b may be used to calculate the sensible cooling load for infiltration air. The latent cooling load is a result of the difference in moisture content between the indoor and outdoor air. The following equation may be used to calculate the latent cooling load for infiltration air:

$$q = (4,840)(cfm)(W)$$

Equation 8.5

NAMEPLATE OR BRAKE HORSEPOWER	FULL LOAD MOTOR EFFICIENCY PERCENT	LOCATION OF EQUIPMENT WITH RESPECT TO CONDITIONED SPACE OR AIR STREAM**		
		Motor In - Driven Machine In $\dfrac{\text{HP} \times 2545}{\% \text{ Eff}}$	Motor Out - Driven Machine In HP x 2545	Motor In - Driven Machine out $\dfrac{\text{HP} \times 2545 (1 - \% \text{ Eff})}{\% \text{ Eff}}$
		BTU per Hour		
1/20	40	320	130	190
1/12	49	430	210	220
1/8	55	580	320	260
1/6	60	710	430	280
1/4	64	1,000	640	360
1/3	66	1,290	850	440
1/2	70	1,820	1,280	540
3/4	72	2,680	1,930	750
1	79	3,200	2,540	680
1 1/2	80	4,770	3,820	950
2	80	6,380	5,100	1,280
3	81	9,450	7,650	1,800
5	82	15,600	12,800	2,800
7 1/2	85	22,500	19,100	3,400
10	85	30,000	25,500	4,500
15	86	44,500	38,200	6,300
20	87	58,500	51,000	7,500
25	88	72,400	63,600	8,800
30	89	85,800	76,400	9,400
40	89	115,000	102,000	13,000
50	89	143,000	127,000	16,000
60	89	172,000	153,000	19,000
75	90	212,000	191,000	21,000
100	90	284,000	255,000	29,000
125	90	354,000	318,000	36,000
150	91	420,000	382,000	38,000
200	91	560,000	510,000	50,000
250	91	700,000	636,000	64,000

*For intermittent operation, an appropriate usage factor should be used, preferably measured.

If motors are overloaded and amount of overloading is unknown, multiply the above heat gain factors by the following maximum service factors:

Maximum Service Factors

Horsepower	1/20 - 1/8	1/6 - 1/3	1/2 - 3/4	1	1 1/2-2	3 - 250
AC Open Type	1.4	1.35	1.25	1.25	1.20	1.15
DC Open Type	--	--	--	1.15	1.15	1.15

No overload is available with enclosed motors.

**For a fan or pump in air conditioned space, exhausting air and pumping fluid to outside of space, use values in last column

Table 8-11. Heat gain from electric motors. (Courtesy, Carrier Corporation, Copyright 1965, Mc-Graw Hill, Inc. Used by permission.)

where:
  q = latent cooling load (Btuh)
  4,840 = a conversion factor for specific heat, mass flow, and time
  cfm = infiltration airflow rate (cfm)
  W = difference in humidity ratio between indoor and outdoor air in lb of water vapor per lb of dry air

Although charts are available that give humidity ratios (W) for air at various dry bulb and wet bulb temperatures, it is more convenient to obtain these values from a psychrometric chart. A complete discussion of the psychrometric chart is included in Chapter 9, along with ventilation cooling load calculations.

Example 8-3.

A room is to be maintained at 75°F dry bulb and 50% relative humidity (rh). The outdoor air conditions are 94°F dry bulb and 75°F wet bulb. If 100 cfm of outside air is infiltrating into the room, what are the sensible, latent, and total cooling loads from the infiltration?

From Equation 6.1b, the sensible load may be calculated directly:

$$q_s = (1.10)(100 \text{ cfm})(94°F - 75°F) = 2,090 \text{ Btuh}$$

From a psychrometric chart (see Chapter 9), the humidity ratios are as follows:

Inside; W = 0.0092 lb water/lb dry air
Outside; W = 0.0144 lb water/lb dry air

Using Equation 8.5, the latent load is:

$$q_i = (4,840)(100 \text{ cfm})(0.0144 - 0.0092) = 2,517 \text{ Btuh}$$

The total load is:

$$q_t = 2,090 + 2,517 = 4,607 \text{ Btuh}$$

System cooling loads

System heat gains, or cooling loads, are internal cooling loads that apply to an entire air conditioning system as opposed to an individual space. Examples of system loads are air-handling unit fan motor heat gains, return-air plenum gains, and duct heat gains.

All of these are loads on an entire system and not on an individual space. The minimum amount of cooling air supplied to a space in order to satisfy the space load should be based on the cooling load for that space (building codes may also set minimum air quantities). Therefore, it is necessary to distinguish between system (or return) loads and space (or room) loads.

Fan-motor cooling loads

A fan and motor add energy to an air conditioning system in the form of pressure and motion of the air (see Chapter 2). Eventually all of the energy supplied to the circulated air becomes heat due to friction losses as the air flows through the ductwork. In order to properly size an air conditioning system, it is necessary to account for this heat gain, particularly in large systems.

If the fan motor is in the airstream (or if the fan and motor are in the conditioned space), the values in Table 8-11 for "Motor In - Driven Machine In" should be used. If the fan and motor are outside the conditioned space and the motor is not in the circulated airstream, the "Motor Out - Driven Machine In" values should be used.

Since a fan is not selected and the fan motor is not sized until after the cooling loads have been determined, it is usually necessary to estimate the size of the fan and motor for cooling load purposes. After the actual fan and motor have been selected, the calculations can be adjusted to reflect the actual values.

Fan brake horsepower is the power supplied directly to the fan itself. It does not take losses in motor efficiency or fan-motor drive losses into account. In order to get a ballpark estimate of a system's fan-motor brake horsepower, use the following equation:

$$Bhp = \frac{QP_s}{6,356 \, N_f}$$

Equation 8.6

where:
- Bhp = brake horsepower supplied to the fan (hp)
- Q = system airflow (cfm)
- P_s = system static pressure (in. wc)
- 6,356 = a unit conversion factor
- N_f = fan static efficiency

It is necessary to do a preliminary estimate for most of the variables in Equation 8.6. For a reasonably good ventilation rate in occupied spaces, ASHRAE recommends 4 to 12 air changes per hour for most spaces. The system airflow may then be approximated as follows:

$$Q = (C_a)(V)\left(\frac{1 \text{ min}}{60 \text{ sec}}\right)$$

Equation 8.7

where:
C_a = space air changes per hour
V = total space volume (cu ft)
60 = hours-to-minutes conversion

Unless a system is specifically designed otherwise, most air systems are low pressure, low velocity (less than 2 in. wc and less than 2,000 fpm). A value of 2 in. wc may be assumed for P_s in Equation 8.6. The fan static efficiency (N_f) can vary widely, depending on the type and arrangement of the fan. A value of 50% is a reasonable average value to assume.

Since all of the heat gain is absorbed by the circulated airstream, there is no thermal storage; therefore, no CLTDs or CLFs need to be used. The values from Table 8-11 may be used directly for system cooling loads.

Example 8-4.
An office has a total volume of 100,000 cu ft. Assume a low-velocity, low-pressure system with the fan motor outside the airstream. The fan is located in an unconditioned penthouse mechanical room. What is the approximate system cooling load for the fan and motor?

Equation 8.7 may be used to estimate the system airflow. Assume 8 air changes per hour:

$$Q = (8)\ (100{,}000 \text{ cu ft})\left(\frac{1 \text{ min}}{60 \text{ sec}}\right) = 13{,}333 \text{ cfm}$$

A fan static pressure of 2 in. wc and a fan static efficiency of 50% is assumed. Equation 8.6 may be used to calculate the fan-motor brake horsepower:

$$Bhp = \frac{(13{,}333 \text{ cfm})\ (2 \text{ in. wc})}{(6{,}356)\ (0.5)} = 8.4 \text{ hp}$$

Assume also a 10-hp motor, which is the next largest standard size motor. Since the air-handling unit is in a penthouse mechanical room and the motor is not in the airstream, the fan is in the conditioned space (the airstream) and the motor is out (Motor Out - Driven Machine In). From Table 8-11, the heat gain for the fan and motor is 25,500 Btuh, which is the cooling load.

Return-air plenum cooling loads

Frequently, the space above a ceiling is used as a return-air plenum. That is, the air is returned to the air-handling system through the ceiling instead of through return-air ducts. In this situation, the return air absorbs heat from the plenum as it is returned to the air-handling unit. These plenum loads should be considered as system or return loads instead of cooling loads on the conditioned space.

A typical example in which part of a heat gain should be considered a plenum load is light fixtures in a ceiling with a return-air plenum. Referring to Figures 8-1 and 8-2, 50% of the heat gain for unvented fixtures is directed up into the plenum, with the balance directed to the space below. For vented light fixtures, 60% of the heat goes up to the plenum, with 40% to the space below. When calculating cooling loads from lights, the effect of system or plenum loads should be taken into account.

Direct heat gains

Whenever supply-air ductwork is routed through an unvented ceiling space (not a plenum) or when ductwork passes through an unconditioned space, the air in the duct may gain sensible heat by conduction. The amount of heat gain is proportional to the duct surface area, the overall heat transfer coefficient (U value) of the duct insulation, and the temperature difference between the air inside the ductwork and the surrounding air. Overall heat transfer coefficients for ductwork and insulation are given in Table 8-12.

Description of ductwork	U value
Sheet metal, not insulated	1.18
Insulation board with or without sheet metal	
1/2 in	0.38
1 in.	0.22
1-1/2 in.	0.15
2 in.	0.12

Table 8-12. Overall heat transfer coefficient for ductwork.

Equation 4.1 may then be used to calculate the sensible heat gain to the air in the ductwork. No CLTDs or CLFs need to be applied.

Unless calculations are being made for an existing building with an existing system, it is difficult to estimate the ductwork for a new system. As a rule of thumb, a sensible heat gain of 1% of the space's sensible cooling load may be assumed for duct heat gains.

Internal cooling load calculation procedure

In Chapter 7, the need for experience and judgment when calculating cooling loads was discussed. Methods for identifying the dominant cooling loads for a space, and when they occur, were also discussed. In addition to reviewing the external loads to see which loads may be most dominant, the internal loads must also be taken into account. In some buildings, such as office buildings, the lighting and equipment loads may be the dominant loads, or they may at least have a significant effect on when the peak will occur. In some buildings like theaters or sports arenas, people may be the dominant load.

All cooling loads (internal and external) must be examined to determine when the space and/or system peak load might occur. Cooling load calculations for both interior and exterior loads should be made for the same times or hours. In order to help organize the calculations, an Internal Cooling Load Calculation form has been provided. This form is to be used in conjunction with the External Cooling Load Calculation form in Chapter 7.

The procedure for estimating the various internal cooling loads and combining the internal loads with the external loads is as follows:

1. Identify the lighting type, electrical power input (and/or output), and the number of fixtures of each type. Calculate the total wattage for the lighting.

2. Based on the loads judged to be the dominant loads (internal and external), select the times for which calculations will be made. Determine the total hours that the lights will be on.

3. Identify the lighting fixture type and estimate the ventilation rate for the space (low, medium, or high). Based on the construction type, the ventilation rate, and the fixture type, select an "a" value from Table 8-3 and a "b" classification from Table 8-4.

4. Based on the "a" value and the "b" classification, the total time the lights are on, and the hours corresponding to the calculation times, select CLFs for each calculation time.

5. Multiply the lighting wattage by a ballast factor (if any). Note that the wattage rating for some fixtures includes the ballast. Multiply the total by 3.413 to convert from watts to Btuh. This will give the instantaneous heat gain for the lights.

6. Determine what percentage of the lighting heat gain is a return gain (if any) and what percentage is a space heat gain.

7. Using the space lighting heat gain, calculate the lighting cooling load for each hour by multiplying the room lighting heat gain by the respective CLFs.

8. Enter the room lighting cooling loads for each calculation hour in the three right hand columns on the form. Repeat the procedure for each type of lighting fixture.

9. Estimate the number of occupants in the space for each calculation hour. Separate into groups according to activity level. Also group by sex and age, if known.

10. Obtain a sensible and latent heat gain for each activity level group from Table 8-1.

11. Apply any age or sex adjustment, if necessary.

12. Estimate the total number of hours the occupants will be in the space and how long it has been since they entered the space for each calculation hour, for each group. Select the appropriate CLFs for the calculation hours from Table 8-2.

13. To determine the sensible cooling load, multiply the number of occupants in each group by the sensible heat gain per occupant. Multiply the total for each group by the appropriate CLF for each hour to get the cooling load for each calculation time.

14. To get the latent cooling load, multiply the number of occupants in each group by the latent heat gain per occupant. The value should be the same for each calculation time if the occupancy and activity level remains constant for the time period.

15. Obtain the sensible and latent (if any) heat gains for each type and piece of equipment from the nameplate or manu-facturer. If this information is not available use the typical values listed in Tables 8-8, 8-9, 8-10, and/or 8-11.

16. To obtain the sensible cooling loads, determine how long the equipment will be on and how long the equipment has been on for each calculation hour. Select an appropriate CLF from Tables 8-6 and/or 8-7. Multiply the equipment heat rejection rate by the watts-to-Btuh conversion (if necessary) and by the CLF for each calculation hour.

17. Based on whether the equipment is hooded or not, the latent equipment gains may be multiplied to get the latent cooling load for the space. The value should be the same for each hour unless

equipment is turned on and off during the calculation period.

18. Estimate the space infiltration rate in accordance with the methods in Chapter 5. Ventilation air should be handled separately (as will be discussed in Chapter 9).

19. Obtain the outdoor design conditions from Table 3-1.

20. Use Equation 6.1b and the indoor-outdoor dry bulb temperature difference to calculate the sensible cooling load. The load should be the same for each hour unless there are significant changes in outdoor temperature during the calcu-lation period.

21. Plot the indoor design conditions and outdoor design conditions on a psychro-metric chart to get the indoor and out-door humidity ratios. Refer to Chapter 9 for a discussion on psychrometric charts. Use Equation 8.5 to calculate the latent loads. They are also normally the same for each calculation hour.

22. Add all of the internal sensible cooling loads to get an internal subtotal, and add all of the internal latent loads to get an internal latent cooling load subtotal.

23. Bring the external sensible cooling load subtotal and the external latent subtotal forward for each calculation time. Add the external subtotals to the internal subtotals to get the space cooling load subtotals. These values may be used to determine the minimum air quantities supplied to the space in order to satisfy the space cooling load.

24. Estimate any system or return cooling loads in accordance with the section on fan-motor cooling loads and Step 6 above.

25. Add the system or return loads to the space cooling loads to get the Building Sensible Cooling Load (BSCL) and the Building Latent Cooling Load (BLCL) for each calculation hour. The hour that has the greatest total cooling load is when the peak load occurs, and the total

is the peak building or space cooling load. These values, along with the ventilation loads, will be combined to get the grand total loads (see Chapter 9).

This procedure is outlined in the table on the following page.

INTERNAL AND TOTAL COOLING LOADS

INTERNAL SENSIBLE

LIGHTS	TYPE	TOTAL HOURS ON	START TIME	WATTS TO BTUH	BALLAST AND USE FACTOR	TOTAL GAIN	CLF		
							__HR	__HR	__HR
				3.413					
				3.413					
				3.413					

PEOPLE	HOURS IN SPACE	TIME ENTER	NUM. OF PEOPLE		GAIN/ PERSON	NUM. x GAIN	CLF		

EQUIP	TYPE		HRS. ON	START TIME	HOODED YES/NO	SENSIBLE GAIN	CLF		

MISC				

COOLING LOAD

___HR	___HR	___HR

INFILTRATION—SENSIBLE	TEMP DIFF		
SCFM X 1.10 X T.D.			

INTERNAL SENSIBLE SUBTOTAL LOAD

EXTERNAL SUBTOTAL COOLING LOAD

SPACE SENSIBLE SUBTOTAL LOAD

INTERNAL LATENT LOAD	GAIN/ PERSON	__HR	__HR	__HR
PEOPLE				
INFILTRATION SCFM x 4840 x W. DIFF.				
EQUIPMENT				
		SUBTOTAL		

OASH: SCFM X 1.10 X TEMP. DIFF.

OALH: SCFM X 4840 X W. DIFF.

GRAND TOTAL COOLING LOAD

Chapter 9

Hvac Psychrometrics

Air conditioning involves the control of both the temperature and moisture content of the air in a conditioned space. The space temperature and moisture content are controlled by various thermodynamic processes provided by the hvac system. Psychrometrics is the study of the thermodynamic processes of air and the moisture that air contains.

Proper control of the space temperature and humidity has a direct effect on occupant comfort. Chapter 1 discussed some of the requirements for providing a comfortable indoor environment. The total cooling load for an air conditioning system is the sum of the space sensible and latent cooling loads, as well as sensible and latent loads from the introduction of outdoor ventilation air.

Air is actually a mixture of several substances, including gases and vapors. Dry air contains no moisture. Its two main components are gaseous nitrogen and gaseous oxygen. Nitrogen accounts for about 78% by volume and oxygen about 21%. Table 9-1 shows the gases that make up atmospheric air.

Atmospheric air is also comprised of small amounts of water vapor. The average amount of water vapor in air is about 0.08% by weight at 0°F and about 1.56% at 70°F when the air is fully saturated with water vapor. Atmospheric air also contains some particulate matter such as dust and pollen.

Standard atmospheric air is considered to have a temperature of 60°F and a density of 0.075 lb/cu

Gas	Percent by weight	Percent by volume
Nitrogen	75.47	78.03
Oxygen	23.19	20.99
Argon	1.29	0.94
Carbon dioxide	0.05	0.03
Hydrogen		
Xenon, Krypton		
and other gases	0.00	0.01
		minute portions

Table 9-1. Gaseous composition of dry air.

ft of dry air at a pressure of 14.7 psia. Although these values may vary, standard air may be assumed for most hvac calculations.

Since air is a mixture of dry air and water vapor, it is necessary to specify two independent properties in order to define the condition of air for any given pressure. These two properties are the temperature and the moisture content of the air. They are necessary to define the condition or "state point" of the air. Methods for determining the state point of air for various thermodynamic and air conditioning processes will be discussed in the following sections of this chapter.

Air and water vapor mixtures

Because air is composed of a number of gases, it may be treated as an "ideal" or "perfect gas." A perfect gas is one in which the molecules are considered to be perfectly elastic, the volume occupied by the molecules is very small in comparison with the total volume, the attractive forces between adjacent molecules is very small, and the molecules move in random directions.

An ideal gas is also one that follows the **Ideal Gas Law** (an equation of state):

$$PV = MRT$$

Equation 9.1

where:

- P = absolute pressure (psia)
- V = volume (cu ft)
- M = mass (lb)
- R = a gas constant
- T = absolute temperature in degrees Rankine (R)

Vapors, such as water vapor (moisture), do not follow the Ideal Gas Law. This is the main difference between vapors and gases. Although vapors do not follow the Ideal Gas Law, in some cases they come close to acting like ideal gases. Therefore, vapors may usually be treated as ideal gases with a reasonable degree of accuracy.

Partial pressures of mixtures

According to **Dalton's Law of Partial Pressures**, the pressure exerted by each gas in a mixture of gases is called the partial pressure of that gas:

$$P_t = p_a + p_b + p_c \ldots$$

Equation 9.2

where:

- P_t = total pressure of the mixture
- p_a = partial pressure exerted by gas A
- p_b = partial pressure exerted by gas B
- p_c = partial pressure exerted by gas C, and so on

The total pressure of an air-vapor mixture is the sum of the partial pressures of each gas and the water vapor. The total barometric pressure of an air-vapor mixture can be expressed as follows:

$$P_b = p_a + p_v$$

Equation 9.3

where:

- P_b = total barometric pressure
- p_a = partial pressure of dry air
- p_v = partial pressure of the water vapor in the mixture

Evaporation

All liquids have a tendency to evaporate. Some liquids, such as rubbing alcohol or gasoline, evaporate quickly. Other liquids, such as mercury, evaporate extremely slowly.

A liquid evaporates because there is always some of the liquid present in the air immediately above it as a vapor. This vapor, which is in contact with the liquid, has a pressure. This pressure is the equilibrium vapor pressure, or vapor pressure. When the partial pressure of the vapor and liquid are equal, evaporation does not take place. The number of molecules leaving the liquid and entering the vapor is equal to the number leaving the vapor and entering the liquid.

Evaporation occurs when some of the vapor diffuses or is blown away. When this happens, more of the liquid will turn to vapor in order to maintain the partial pressure of the vapor above the liquid. Evaporation continues until the partial pressure of the vapor in the air is equal to the pressure of the liquid water. At this point evaporation will cease, the partial pressure of the vapor and the pressure of the water will be in equilibrium, and the air-vapor mixture will be "saturated."

The maximum amount of water vapor that air can hold in suspension at any given pressure is a function of its dry bulb temperature. The air-vapor mixture is said to be saturated when it can no longer hold any additional moisture at that temperature. The moisture that is in suspension is on the verge of condensing out.

The evaporation of water requires latent heat. As a result of the evaporation of water into the air, water tends to cool as it surrenders the necessary heat for evaporation. This heat is called the **latent heat of vaporization**.

Condensation of water vapor

The condensation of water from an air-vapor mixture requires the same amount of latent heat as evaporation for the same set of conditions. The main differences are that the net flow of water molecules is out of the air-vapor mixture and the flow of latent heat is out of the air-vapor mixture.

In order for condensation to occur, the air-vapor mixture must be cooled to the saturation point. The saturation point occurs at the saturation temperature of the mixture. The saturation temperature (or dewpoint temperature) is the temperature at which the vapor begins to condense.

When an air-vapor mixture comes in contact with a cool surface, the vapor in the air will have a higher vapor pressure than the equivalent vapor pressure for the temperature of the cooler surface. This will result in a flow of water vapor molecules to the surface. As more molecules flow to the surface, they will combine and water droplets will begin to form on the surface.

The sole criterion as to whether evaporation or condensation will take place is the relation between the dewpoint temperature of the air-vapor mixture and the temperature of the water on a surface in contact with the air mixture.

Relative humidity

Humidity is the moisture or water vapor occupying the same space as the air in an air-vapor mixture. Absolute humidity or humidity ratio (W) is the weight of the water vapor per unit volume of dry air. It is usually expressed in pounds of moisture per pound of dry air. There are 7,000 grains of moisture in a pound of water.

The relative humidity (rh) of an air-vapor mixture is defined as the ratio of the pressure of the vapor in the mixture to the saturation pressure of the vapor at the same temperature or the temperature of the mixture. Rh is expressed as a percentage and is expressed as follows:

$$\phi = \frac{P_v}{P_g}$$

Equation 9.4

where:

P_v = pressure of the vapor in the air-vapor mixture

P_g = saturation pressure of the vapor

Rh is an important measure of the amount of moisture in an air-vapor mixture for any given temperature. It is the most commonly used measurement for expressing the moisture content in an air-vapor mixture. The higher the percent rh, the higher the moisture content at a given temperature. If the temperature of the mixture is increased with no change in the moisture content of the mixture, rh will decrease. This can be seen graphically on a psychrometric chart and will be discussed later in this chapter.

Dry bulb and wet bulb temperature

The dry bulb temperature of an air-vapor mixture is simply the temperature that would be read from an ordinary thermometer. Changes in dry bulb temperature would indicate a change in sensible heat. The wet bulb temperature of an air-vapor mixture is a measure of the humidity level or moisture content of the mixture at a given dry bulb temperature. The dry bulb and wet bulb temperatures define the state or identify a state point of the mixture.

The wet bulb temperature of air is measured using a psychrometer: a thermometer with a thin cloth (wick), wet with distilled water, covering the bulb of the thermometer (hence the term wet bulb temperature). Air is passed over the bulb and wick at a prescribed rate. If the water vapor in the air is not saturated, evaporation takes place, and latent heat is transferred to the moving air. This lowers the temperature of the water in the wick until an equilibrium temperature is reached. The temperature at equilibrium is the wet bulb temperature of the air-vapor mixture.

The difference between the dry bulb and wet bulb temperature is called the **wet bulb depression**. If the air-vapor mixture is at saturation, very little evaporation will take place. At this state, the dry bulb and wet bulb temperatures will be nearly equal. On the other hand, if the moisture in the mixture is low, there will be a high rate of evaporation and a lower wet bulb temperature will be measured.

A common device for measuring dry bulb and wet bulb temperatures is the sling psychrometer shown in Figure 9-1. The sling psychrometer has two thermometers, one for measuring dry bulb temperature and the other for measuring wet bulb temperature.

The thermometer for measuring the wet bulb temperature has a cloth wick on the bulb, which is wetted by dipping it in water. The thermometers are then whirled around to create the necessary air movement. The sling psychrometer is whirled until the two temperature readings remain constant.

Enthalpy of air-vapor mixtures

In Chapter 2, the thermal property of a substance known as enthalpy was discussed. Enthalpy (h) is a measure of the internal energy or total heat content of a substance.

The enthalpy of a mixture of perfect (ideal) gases or of an air-vapor mixture is the sum of the enthalpies of each constituent gas or vapor. Therefore, the enthalpy of an air-vapor mixture is actually the total heat of the mixture. The customary units for an air-vapor mixture are Btu/lb of dry air.

Since enthalpy is a measure of the total heat content of the air-vapor mixture, it is used extensively in analyzing hvac processes. The reason is because these hvac processes add and remove sensible and latent heat to or from air-vapor mixtures.

Psychrometric chart

Psychrometrics can be defined as the science involving the thermodynamic properties of moist air and the effect of atmospheric moisture on materials and human comfort. A psychrometric chart is a plot of the psychrometric properties of moist air. A psychrometric chart is shown in Figure 9-2.

The chart is for the normal temperature range of hvac processes (32° to 105°F dry bulb) and for the standard barometric pressure of 14.7 psia (29.92 inches of mercury, or in. Hg). If the total pressure of the air differs significantly from 14.7 psia, serious errors will result in the use of this particular chart. For other pressures, different charts should be used or corrections should be applied.

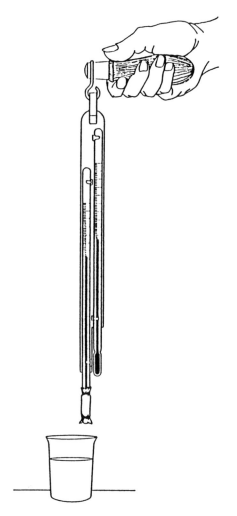

Figure 9-1. Sling psychrometer.

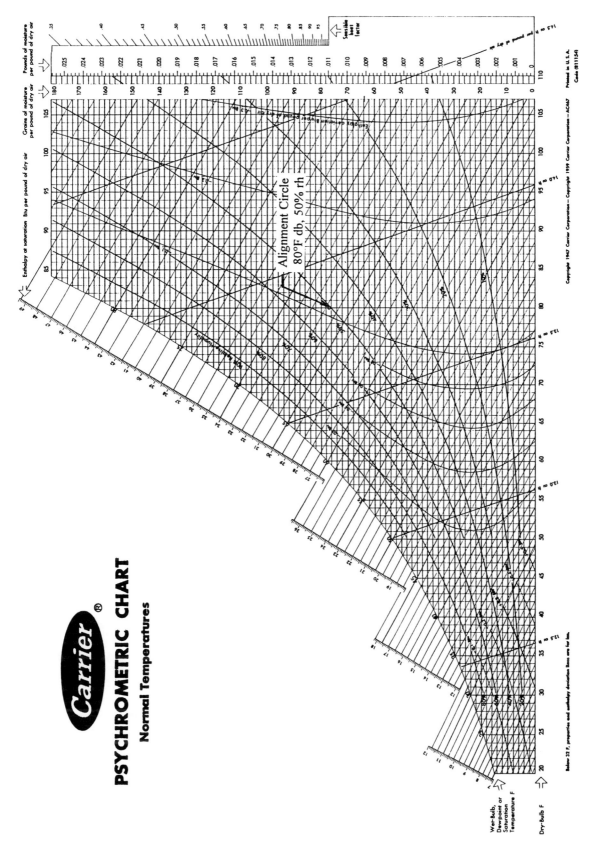

Figure 9-2. Psychrometric chart. (Courtesy, Carrier Corporation)

The chart is also useful in that if two properties of air are known, such as dry bulb and wet bulb temperatures, a state point has been defined and all the other properties for that state point may be obtained. (The state point would be located on the chart at the intersection of the two known properties.)

A skeleton psychrometric chart that indicates where the various properties, scales, and features are located on a "psych" chart is shown in Figure 9-3.

Dry bulb temperature (Line 1): The temperature read on a standard thermometer in degrees Fahrenheit dry bulb (°F db). Lines of constant dry bulb temperature are vertical straight lines on the chart; the temperature scale is across the bottom of the chart.

Wet bulb temperature (Line 2): The temperature read with a psychrometer. Lines of constant wet bulb temperature are straight lines that slope upward to the left. Wet bulb temperatures are given in degrees Fahrenheit wet bulb (°F wb) and are read along the curve on the upper left side of the chart.

Humidity ratio, W (Line 3): The ratio indicates the absolute moisture content of the air-vapor mixture in grains of moisture per pound of dry air, or in Btu per pound of dry air. Lines of constant humidity ratio, or absolute moisture content, are horizontal lines. The humidity ratio is read to the right side of the chart.

Rh (Line 4): Lines of constant rh are the lines inside the chart that curve upward to the right.

Specific volume (Line 5): The volume of the air-vapor mixture per unit weight given in cubic feet per pound of dry air. The lines slope diagonally upward to the left.

Saturation curve (Line 6): The saturation curve, or dewpoint, is a line at which the dry bulb and wet bulb temperatures are nearly equal, resulting in a saturated air-vapor mixture. An air-vapor mixture must be cooled (have its dry bulb temperature lowered) to this point before any significant condensation will occur.

Enthalpy scale (Scale 7): Lines of constant enthalpy of the air-vapor mixture.

Sensible heat ratio (Scale 8): The sensible heat ratio (SHR) or sensible heat factor (SHF) is the ratio of the sensible heat to the total (sensible and latent) heat in an hvac process. SHR is read from the scale to the right of the chart.

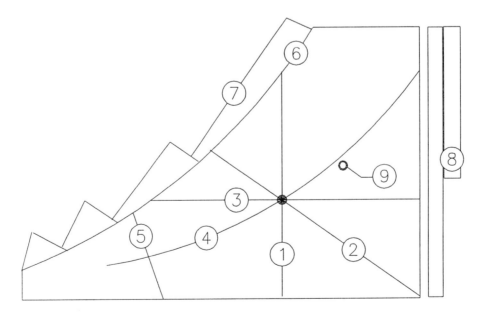

Figure 9-3. Skeleton psychrometric chart.

Alignment circle (Point 9): A point located inside the chart used in conjunction with SHR to plot various psychrometric processes. In Figure 9-2, the alignment circle (or point) is located at 80°F db and 50% rh. Some other charts locate the alignment circle at 78°F db and 50% rh.

Hvac processes

Figure 9-4 shows the typical processes used in hvac systems on a skeleton psychrometric chart. The lines show the variation in properties that the air-vapor mixture would pass through as it changes from one state point to another. State point 0 is assumed to be the beginning point of the process.

The processes in actual hvac systems are frequently more complex and involved than those shown in Figure 9-4. However, most hvac processes can be analyzed by combinations of the basic processes. Any change in state point of the air-vapor mixture can be broken down into sensible and latent heat changes. These can be plotted on a psychrometric chart. From this point on, the air-vapor mixture will just be referred to as air. Referring to Figure 9-4, the basic processes may be observed.

Humidification only (state point O to A) and dehumidification (state point O to E): These processes involve an increase (O to A) or a decrease (O to E) in the humidity ratio, with no change in the dry bulb temperature. These are purely changes in latent heat, with no change in sensible heat content.

Sensible heating only (state point O to C) and sensible cooling only (state point O to G): These processes involve an increase (O to C) and a decrease (O to G) in the dry bulb temperature of the air. These are purely changes in sensible heat, with no change in the moisture content. The humidity ratio (W) remains constant.

Combined cooling and dehumidification (state point O to F): This is a result of the reduction in both dry bulb and wet bulb temperatures. Cooling coils normally perform a process like this, where both sensible and latent heat of the air decreases.

Combined heating and humidification (state point O to B): This is a result of an increase in both dry bulb and wet bulb temperatures. In this process, both sensible and latent heat are added to the air.

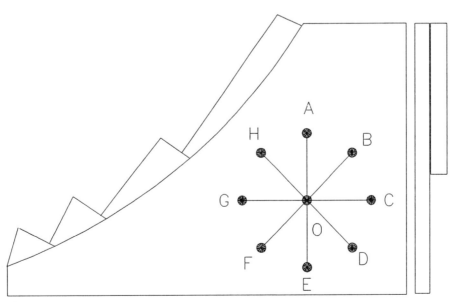

Figure 9-4. Basic hvac processes.

Evaporative cooling only (state point O to H):
This is a process in which cooling and
humidification occur. The wet bulb temperature
of the air remains constant as the dry bulb
temperature is reduced. This is an adiabatic (no
net heat transfer) process. The process removes
sensible heat, with the heat vaporizing an additional amount of water added (added latent heat).

**Chemical dehumidification (state point O to
D):** This is a process in which moisture is
removed from the air chemically. This is a
constant enthalpy process, with the latent cooling
nearly equal to the sensible heat increase.

Figure 9-5 shows a schematic of a typical hvac
system. A typical air conditioning unit includes a
circulating fan with a motor and a cooling coil
through which air is passed. The unit may or
may not include a means of heating the circulated
air. The cooling coil may be a chilled water coil
or a direct expansion (DX) coil with a refrigerant.

The return air (ra) volume is usually less than the
supply air (sa) volume due to space exhaust,
exfiltration out of the building, and relief air at
the unit (air exhausted at the unit in order to
provide an adequate outdoor air ventilation rate).

The outdoor air (oa) volume is the difference
between the sa and ra volume. The oa volume is
usually set by codes (see Chapter 5). Equation
9.5 expresses the relationship between the various
air quantities:

$$Q_{sa} = Q_{ra} + Q_{oa}$$

Equation 9.5

where:
Q_{sa} = supply air volume (cfm)
Q_{ra} = return air volume (cfm)
Q_{oa} = outdoor air volume (cfm)

The minimum sa volume may be determined by
the methods described later in this chapter or by
code requirements.

The room sensible heat factor (RSHF) is the ratio
of the room sensible heat (cooling load) to the
total room heat (sensible plus latent). Equation
9.6 is the equation for RSHF:

$$RHSF = \frac{Q_{rs}}{Q_{rs} + Q_{rl}} = \frac{Q_{rs}}{Q_{rt}}$$

Equation 9.6

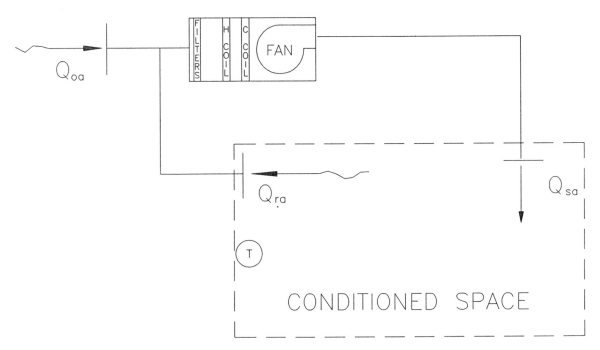

Figure 9-5. Schematic of a typical hvac system.

where:

Q_{rs} = room sensible heat or cooling load (Btuh)

Q_{rl} = room latent heat or cooling load (Btuh)

Q_{rt} = room total heat or cooling load (Btuh)

When plotted on a psychrometric chart, the slope of the RSHF line graphically shows the ratio of room sensible load to the room total load. A typical sensible heat factor line is shown between state points 3 and 4 in Figure 9-6.

When outdoor air is brought into an hvac system, the system must remove the cooling loads due to the outdoor air as well as the space cooling loads. The grand total cooling load is the sum of the space (sensible and latent) cooling loads, as well as the outdoor (sensible and latent) loads. The grand total sensible heat factor (GSHF) is the ratio of the total sensible heat (space plus outdoor) cooling load to the grand total cooling load. Equation 9.7 is the equation for GSHF:

$$GSHF = \frac{Q_{st}}{Q_{st} + Q_{lt}} = \frac{Q_{st}}{Q_{gt}}$$

Equation 9.7

where:

Q_{st} = total sensible load (Btuh)

Q_{lt} = total latent load (Btuh)

Q_{gt} = grand total cooling load (Btuh)

When plotted on a psychrometric chart, the slope of the GSHF line graphically shows the ratio of the total system sensible loads to the grand total cooling loads. A typical GSHF line is shown between state points 2 and 3 in Figure 9-6.

Figure 9-6 shows the air state points usually found in an air cooling and dehumidifying process. As the conditions of the air pass from one state point to another, it is assumed that the process is really a series of state points that make up the interconnecting lines.

Ambient outside air conditions (state point 1): This point is located, or described, by the design dry bulb and wet bulb temperatures for the location of the building. The point is located by the intersection of the dry bulb and wet bulb temperature lines. These values are obtained from climatic data (see Chapter 3).

Mixed-air conditions (state point 2): The conditions at this point are a function of the outdoor air conditions, the return air conditions, and the quantities of outside and return air. The

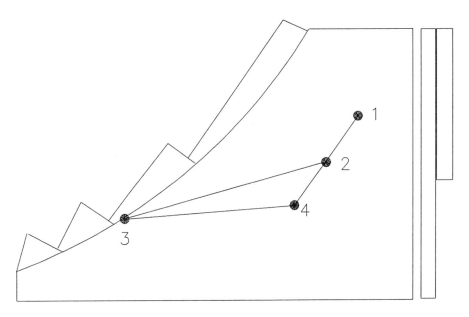

Figure 9-6. Typical cooling processes.

dry bulb temperature of the mixed air may be calculated using the following equation:

$$T_{mdb} = \frac{Q_{oa}T_{oa} + Q_{ra}T_{ra}}{Q_{oa} + Q_{ra}}$$

Equation 9.8

where:

T_{mdb} = mixed-air temperature (°F db)
Q_{oa} = outside airflow rate (cfm)
T_{oa} = outside air dry bulb temperature (°F db)
Q_{ra} = return airflow rate (cfm)
T_{ra} = return air dry bulb temperature (°F db)

It is usually necessary to estimate the return air temperature in Equation 9.8. Since the hvac system should maintain the room design conditions in the space, the room design temperature may be assumed as a reasonable value of the return air temperature. If there are unusually large return cooling loads, it may be necessary to calculate a return air temperature based on the return airflow rate and return cooling loads.

Since it is known that the mixed-air conditions are between the space design conditions and the outdoor air conditions, the mixed-air state point must be on a line between the indoor and outdoor conditions. By locating the T_{mdb} value on the line between state points 1 and 4, state point 2 is defined and the mixed-air wet bulb temperature (T_{mwb}) may be read from the psychrometric chart.

Leaving air conditions (state point 3): This point is (very nearly) the condition of the air as it comes out of the cooling coil. Notice that the point is at, or near, the saturation (dew) point. In order for any significant dehumidification to occur, the air must be cooled near the dewpoint temperature. As the air is cooled, it is cooled along the line between the mixed-air conditions (state point 3). The slope of this line is parallel to the grand sensible heat factor (GSHF) of the mixed air.

When the conditioned air is supplied to the space, it is usually considered to be at, or near,

the leaving air conditions (state point 3) of the cooling coil. As the air is supplied to the space, it heats up along the line from state point 3 to the room conditions at state point 4. Assuming that adequate air is supplied to handle the cooling loads, the line is parallel to the RSHF. Methods for determining the required leaving air conditions and the required airflow rate are discussed later in this chapter.

Space design conditions (state point 4): These are the design conditions that the hvac system will be designed to maintain within the conditioned space. These are usually a design dry bulb temperature and rh. The point is plotted at the intersection of the dry bulb temperature and the percent rh line.

Example 9-1.

An air conditioning system is serving a space with a room sensible cooling load of 75,000 Btuh and latent cooling load of 5,000 Btuh. Outdoor air is brought into the system, resulting in a sensible cooling load of 7,500 Btuh and a latent cooling load of 4,000 Btuh. What is the RSHF and the GSHF?

From Equations 9.6 and 9.7:

$$RSHF = \frac{75,000 \ Btuh}{75,000 \ Btuh + 5,000 \ Btuh} = 0.94$$

$$GSHF = \frac{75,000 \ Btuh + 7,500 \ Btuh}{75,000 \ Btuh + 7,500 \ Btuh + 5,000 \ Btuh + 4,000 \ Btuh} = 0.90$$

Example 9-2.

An air conditioning system is supplying 10,000 cfm of air to a space that is to be maintained at 75°F db and 50% rh. Outside air is brought into the system at 95°F db and 76°F wb at a rate of 2,000 cfm. What are the conditions entering the cooling coil (the mixed-air conditions), and what is the humidity ratio at each state point?

The outdoor and indoor air state points are plotted on a psychrometric chart, and a straight line is drawn between the two points, Figure 9-7.

Equation 9.5 is used to calculate the return airflow as follows:

$$Q_{ra} = 10,000 \text{ cfm} - 2,000 \text{ cfm} = 8,000 \text{ cfm}$$

Using Equation 9.8, the mixed-air dry bulb temperature may be determined as follows:

$$T_{mdb} = \frac{(2,000 \text{ cfm})(95°F) + (8,000 \text{ cfm})(75°F)}{10,000 \text{ cfm}} = 79°F \text{ db}$$

The mixed-air wet bulb temperature may be determined by plotting 79°F db on the line between the indoor and outdoor state points, as shown in Figure 9-7.

From the psychrometric chart, the entering wet bulb temperature may be read as 65.5°F wb (approximate). The humidity ratios for the state points may also be read directly from the right side of the psychrometric chart, Figure 9-7:

Outdoor: W = 0.0150 lb water/lb dry air
Mixed air: W = 0.0106 lb water/lb dry air
Space air: W = 0.0093 lb water/lb dry air

Applying psychrometrics to cooling load calculations

The final system cooling load calculations involve estimating how much airflow, in dry air, will handle the space cooling and the outdoor air cooling loads. The calculations also involve calculating the conditions, or state points, of the mixed airstream and the conditions leaving the cooling coil. All of this information is necessary to accurately size, select, and specify air conditioning equipment.

The requirements for determining the minimum outdoor air ventilation rate are usually determined by code requirements as discussed in Chapter 5. Methods for calculating the space (room) sensible cooling load, space latent cooling load, return air sensible cooling loads, building sensible cooling load, and building latent cooling load were discussed in Chapters 7 and 8.

Chapter 8 presented methods for calculating cooling loads for outdoor air entering a conditioned space by infiltration. The same methods may be used to calculate cooling loads for outdoor ventilation air brought into an hvac system.

Outdoor air cooling load

The minimum amount of outdoor ventilation air is usually set by code requirements. In some cases, such as locker rooms or kitchens, etc., the minimum outside ventilation air brought in by the hvac system is determined by the amount of air exhausted from the space served by the hvac system. The amount of outdoor air brought in by the hvac system should almost always be equal to or greater than the exhaust air quantity for the space. Negative pressures in buildings should be avoided to reduce infiltration and other operating problems.

The loads imposed by the introduction of unconditioned outdoor air into the space were discussed in Chapters 6 and 8. Just as infiltration loads must be handled by the hvac system, so must the outdoor air ventilation loads. The main difference is that the infiltration heat gains are loads on the conditioned space, while outdoor air loads are loads on the hvac system. This chapter discusses the effect of outdoor air loads on the system.

Equation 6.1b may be used to calculate the outdoor air sensible heat:

$$q_{oas} = (1.10)(Q_{oa})(T_{oa} - T_r)$$

Equation 6.1b

where:

q_{oas} = outside air sensible heat (Btuh)
1.10 = conversion factor
Q_{oa} = outdoor airflow (cfm)
T_{oa} = outside air temperature (°F db)
T_r = room or space air temperature (°F db)

Equation 8.5 may be used to calculate the outdoor air latent heat:

$$q_{oal} = (4,840)(Q_{oa})(W_{oa} - W_r)$$

Equation 8.5

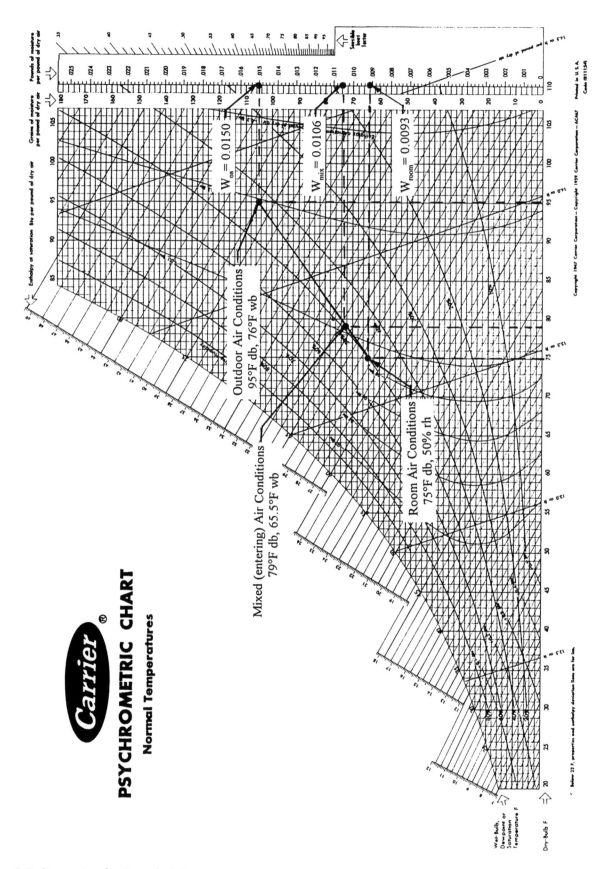

Figure 9-7. State points for Example 9-2.

where:

q_{oal} = outside air latent heat (Btuh)
4,840 = conversion factor
Q_{oa} = outdoor airflow (cfm)
W_{oa} = outside air humidity ratio (lb water/ lb dry air)
W_r = room or space humidity ratio (lb water/lb dry air)

The grand total cooling load (GTCL) is the sum of the space loads, return loads, and outdoor air loads:

$$q_{gt} = q_{rs} + q_{rl} + q_{ras} + q_{oas} + q_{oal}$$

Equation 9.9

where:

q_{gt} = grand total cooling load (Btuh)
q_{rs} = room or space sensible cooling load (Btuh)
q_{rl} = room or space latent cooling load (Btuh)
q_{ras} = return air sensible cooling load (Btuh)
q_{oas} = outside air sensible cooling load (Btuh)
q_{oal} = outside air latent cooling load (Btuh)

Bypass factor and effective sensible heat factor

Whenever air is passed through a cooling coil, most of the air is cooled and usually dehumidified. It is inevitable, however, that some of the air passing through the coil will not be affected. The amount of air that passes through the coil unchanged is called **bypassed air**; the fraction of the total airflow bypassed is called the **bypass factor**. Typical bypass factors for various cooling applications are given below.

0.30 to 0.50 — Residential air conditioners, where cooling loads are not great and cooling is purely for comfort.

0.20 to 0.30 — Small commercial applications, usually using small packaged air conditioning units.

0.10 to 0.20 — Medium to large packaged air conditioning systems, such as packaged rooftop air conditioning units with standard cooling coils.

0.10 to 0.05— Built-up air conditioning units with multiple row cooling coils specifically selected for the application. These are used when fairly precise control of space temperature and humidity are required and for moderately high latent cooling loads.

0 to 0.10 — Built-up air conditioning units with many row cooling coils with close fin spacing. Used for process air conditioning, high latent cooling loads, and/or all outdoor air.

Cooling coils that have many rows and close fin spacing have a larger surface area for heat transfer from the flowing air to the cooling coil. These types of coils tend to have low bypass factors. Coils with comparatively wide fin spacing and few rows tend to have higher bypass factors.

The velocity of the air passing through the coil also affects the bypass factor. Typical air face velocities through cooling coils range from 300 to 800 fpm. Higher face velocities tend to result in higher bypass factors; lower velocities result in lower bypass factors. The amount of coil surface area has a greater effect on bypass factors than does the face velocity.

The actual surface temperature of a cooling coil with chilled water or with refrigerant is not uniform, but varies with the location on the cooling coil. On a multi-row cooling coil, the surface nearest the entering air is in contact with the warmest air and therefore warmer than the average or mean surface temperature. The coil surface near the discharge side of the coil is in contact with air that already has been cooled to some extent. The coil surface temperatures near the discharge side of the coil tend to have a lower surface temperature.

In order to analyze the effect of a cooling coil on the air passed through the coil, it is necessary to assume an effective coil surface temperature that would produce the same leaving conditions as the actual non-uniform surface temperatures. For cooling and dehumidifying situations, the effective coil surface temperature is assumed to be where the GSHF line crosses the saturation or dewpoint curve on the psychrometric chart. This point is commonly called the **apparatus dewpoint temperature (adp)**.

Effective sensible heat factor (ESHF) is the ratio of the effective room (space) sensible heat to the effective room total heat. The effective room sensible heat (ERSH) is the room sensible heat (cooling load) plus the outdoor ventilation air sensible heat that is assumed to be bypassed by the cooling coil. Equation 9.10 gives the effective room sensible heat:

$$q_{esh} = q_{rs} + (BF)(q_{oas})$$

Equation 9.10

where:

q_{esh} = effective room sensible heat (Btuh)
q_{rs} = room sensible heat (Btuh)
BF = coil bypass factor
q_{oas} = outside air sensible heat (Btuh)

The effective room latent heat may be found in a similar fashion by modifying Equation 9.10 as follows:

$$q_{erl} = q_{rl} + (BF)(q_{oal})$$

Equation 9.10a

where:

q_{erl} = effective room latent heat (Btuh)
q_{rl} = room latent heat (Btuh)
BF = coil bypass factor
q_{oal} = outside air latent heat (Btuh)

It is assumed that the outdoor air bypassed through the cooling coil becomes a cooling load on the conditioned space. This additional load affects the required cooling capacity of the hvac system by increasing the amount of conditioned air that must be supplied to the space in order to handle the additional load. The ESHF may be calculated using the following equation:

$$ESHF = \frac{q_{esh}}{q_{esh} - q_{ehl}} = \frac{q_{rs} + (BF)(q_{oas})}{q_{rs} + (BF)(q_{oas}) + q_{rl} + (BF)(q_{oal})}$$

Equation 9.11

Example 9-3.

A packaged rooftop air conditioning unit is serving a space that is maintained at 75°F db and 50% rh and has a sensible cooling load of 50,000 Btuh and a latent cooling load of 2,500 Btuh. The unit coil has a typical bypass factor of 0.15. Neglect any return loads. Air is supplied to the space at the rate of 2,300 cfm. Outdoor ventilation air is brought into the system at a rate of 460 cfm at 94°F db, 75°F wb. What is the grand total heat and the ESHF for the space?

Using Equation 6.1b, the outside air sensible heat is:

$$q_{oas} = (1.10)(460 \text{ cfm})(94°F - 75°F) = 9,614 \text{ Btuh}$$

Plotting the room and outdoor air state points on a psychrometric chart, the indoor and outdoor humidity ratios may be determined. Using Equation 8.5, the outside air latent heat is:

$$q_{oal} = (4,840)(460 \text{ cfm})(0.0144 \text{ lb water/lb dry air} - 0.0092 \text{ lb water/lb dry air}) = 11,577 \text{ Btuh}$$

The grand total cooling load may be calculated using Equation 9.9:

$$q_{gt} = 50,000 \text{ Btuh} + 2,500 \text{ Btuh} + 9,614 \text{ Btuh} + 11,577 \text{ Btuh} = 73,691 \text{ Btuh}$$

The packaged rooftop air conditioning unit has a typical bypass factor of 0.15. The ESHF may be calculated using Equation 9.11:

$$ESHF = \frac{50,000 \text{ Btuh} + (0.15)(9,614 \text{ Btuh})}{\left(\begin{array}{c} 50,000 \text{ Btuh} + (0.15)(9,614 \text{ Btuh}) + \\ 2,500 \text{ Btuh} + (0.15)(11,577 \text{ Btuh}) \end{array}\right)} = 0.92$$

The conditioned air supply

The minimum amount of air supplied to a conditioned room or space is usually a function of the sensible cooling load in the space. During periods of low cooling loads, the minimum airflow may be based on ventilation code requirements.

For design purposes, the conditioned air supply to a space is based on the sensible cooling loads for the space, since cooling is usually controlled by a thermostat that senses the dry bulb temperature of the air in the space. To account for the outdoor air that is bypassed through the cooling coil and becomes a cooling load in the space, the conditioned air supply calculations for the space are based on the space's effective sensible heat.

The total airflow for an entire system is determined in a similar fashion. A simplified one-step method is presented in this section as a calculation method for the conditioned air supply. This involves the use of the effective total sensible heat (space and outdoor), the GSHF, coil bypass factor, and apparatus dewpoint temperature. The reader may consult References 1 and 7 for a more-detailed discussion of the calculations for the conditioned air supply.

Since the GSHF is the relationship of the effective total sensible heat to the grand total heat, the GSHF describes how the air will be cooled and dehumidified as it passes through the cooling coil and looses both sensible and latent heat. When plotted on a psychrometric chart, the slope of the GSHF line shows the relationship of sensible heat to total heat.

To determine the required adp conditions, it is necessary to plot the ESHF line on a psychrometric chart. First, the ESHF must be calculated from the space and outside air cooling loads. The value of the ESHF is then plotted on the SHF scale on the right side of the psychrometric chart. A straight line is then drawn between the ESHF on the SHF scale and the alignment circle near the center of the chart.

Referring back to the psychrometric chart in Figure 9-2, the alignment circle is located at 80°F db, 50% rh. A point locating the room air conditions is then plotted on the psychrometric chart. A second line that passes through the room air

state point is then drawn parallel to the first ESHF line. The second line is extended to where it intersects the saturation curve. The point where the second line intersects the saturation curve defines the required adp conditions for the given loads. The adp defines the effective coil surface temperature; the second line through the room's air point and the adp describes how the air absorbs the sensible and latent heat after it enters the space.

The minimum amount of conditioned air that must be supplied to the space is based on the total effective sensible heat. Equation 9.12 may be used to calculate the minimum quantity of air that must be supplied in order to handle the total sensible cooling load:

$$Q_{da} = \frac{q_{esh}}{(1.1)(T_{rdb} - T_{adp})(1 - BF)}$$

Equation 9.12

where:

Q_{da} = rate of dry conditioned air supplied to the space (cfm)

q_{esh} = effective total sensible heat (Btuh)

T_{rdb} = space or room design dry bulb temperature (°F db)

T_{adp} = apparatus dewpoint dry bulb temperature (°F db)

BF = cooling coil bypass factor

Since some of the outside air is bypassed when the air flows through the cooling coil, part of the outside air cooling load becomes a cooling load on the space by increasing the temperature of the air leaving the cooling coil. This outside air brought into the space is similar to outside air infiltration into the space.

Equation 9.12 incorporates this additional load into the calculation with the bypass factor and effective sensible heat. This results in a coil discharge dry bulb temperature slightly higher than the coil's adp temperature. Equation 9.13 may be used to calculate the leaving air dry bulb temperature:

$$T_{ldb} = T_{adp} + (BF)(T_{mdb} - T_{adp})$$

Equation 9.13

where:

T_{ldb} = leaving dry bulb temperature (°F)
T_{adp} = apparatus dewpoint temperature (°F)
BF = coil bypass factor
T_{mdb} = mixed-air dry bulb temperature (°F)

To determine the leaving wet bulb temperature, the leaving dry bulb temperature from Equation 9.13 may be plotted on the GSHF line to determine the coil leaving air state point. The leaving air wet bulb temperature may be read directly from the psychrometric chart.

Example 9-4.

A space served by a built-up air-handling unit with a multi-row cooling coil has a sensible cooling load of 110,000 Btuh and a latent load of 5,500 Btuh. The coil has a bypass factor of 0.10. The space is to be maintained at 75°F db, 50% rh. Outside air is brought in through the system at a rate of 1,000 cfm; the conditions are 94°F db, 75°F wb.

What is the required total cooling capacity of the air conditioning unit? How much air must be supplied by the air-handling unit to satisfy the cooling loads? What is the coil adp temperature? And what are the conditions of the air as it enters and leaves the cooling coil?

The indoor and outdoor state points are first plotted on the psychrometric chart and a straight line drawn between them, as shown in Figure 9-8. The outside air sensible heat is calculated using Equation 6.1b:

$$q_{oas} = (1.10)(1,000 \text{ cfm}) (94°F - 75°F) = 20,900 \text{ Btuh}$$

The indoor and outdoor humidity ratios are read from the psychrometric chart, and Equation 8.5 is used to calculate the latent cooling load:

$$q_{oal} = (4,840)(1,000 \text{ cfm}) (0.0146 \text{ lb water/lb dry air} - 0.0092 \text{ lb water/lb dry air}) = 26,136 \text{ Btuh}$$

(which may be rounded to 26,150 Btuh)

The total required cooling capacity is calculated from Equation 9.9:

$$q_{gt} = 110,000 \text{ Btuh} + 5,500 \text{ Btuh} + 20,900 \text{ Btuh} + 26,150 \text{ Btuh} = 162,550 \text{ Btuh}$$

Since this is a built-up air conditioning unit with a multi-row cooling coil, a bypass factor of 0.10 is assumed. Equation 9.11 is used to calculate the ESHF:

$$ESHF = \frac{110,000 \text{ Btuh} + (0.10)(20,900 \text{ Btuh})}{\left(\begin{array}{c}110,000 \text{ Btuh} + (0.10)(20,900 \text{ Btuh}) \\ + 5,500 \text{ Btuh} + (0.10)(26,150 \text{ Btuh})\end{array}\right)} = 0.93$$

The value of ESHF = 0.93 is then plotted on the Sensible Heat Factor Scale on the right side of the chart, and a line is drawn through the alignment circle. A second line parallel to the first line is drawn from the space design state point to the saturation curve to get the cooling coil adp temperature, Figure 9-8:

$$T_{adp} = 54°F \text{ db (approximate)}$$

The conditioned air supply rate may then be calculated using Equation 9.12:

$$Q_{da} = \frac{110,000 \text{ Btuh} + (0.10)(20,900 \text{ Btuh})}{(1.1)(75°F - 54°F)(1 - 0.10)} = 5,392 \text{ cfm}$$

The mixed (entering) air dry bulb temperature may be estimated from Equations 9.5 and 9.8:

$$Q_{ra} = 5,390 \text{ cfm} - 1,000 \text{ cfm} = 4,390 \text{ cfm}$$

$$T_{mdb} = \frac{(1,000 \text{ cfm})(94°F) + (4,390 \text{ cfm})(75°F)}{5,390 \text{ cfm}} = 78.5°F \text{ db}$$

The entering dry bulb temperature is plotted on the line between the outdoor and indoor air state points, and the entering wet bulb temperature is read from the chart:

$$T_{mwb} = 65.5°F \text{ wb}$$

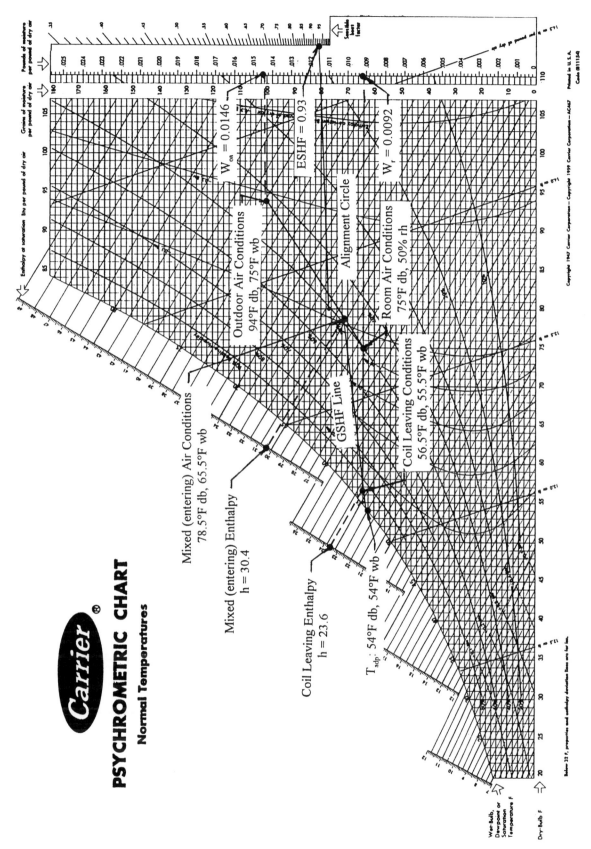

PSYCHROMETRIC CHART
Normal Temperatures

Carrier®

Outdoor Air Conditions
94°F db, 75°F wb

$W_{oa} = 0.0146$

ESHF = 0.93

Alignment Circle

Room Air Conditions
75°F db, 50% rh

$W_r = 0.0092$

Mixed (entering) Air Conditions
78.5°F db, 65.5°F wb

Mixed (entering) Enthalpy
h = 30.4

GSHF Line

Coil Leaving Conditions
56.5°F db, 55.5°F wb

Coil Leaving Enthalpy
h = 23.6

T_{adp}: 54°F db, 54°F wb

Figure 9-8. State points for Example 9-4.

Equation 9.13 may then be used to calculate the leaving air dry bulb temperature:

$$T_{ldb} = (54°) + (0.10)(78.5°F - 54°F) = 56.5°F \ db$$

A straight line is then drawn between the adp conditions and the entering air conditions which represents the GSHF. The leaving dry bulb temperature is then plotted on the GSHF line, and the leaving wet bulb temperature is read from the psychrometric chart:

$$T_{lwb} = 55.5°F \ wb$$

As a check of the calculations, the total cooling capacity and the sensible cooling capacity may be calculated by using the calculated airflow rate, entering air conditions, and leaving air conditions. Equation 9.14 may be used as an alternate method to calculate the total cooling capacity:

$$q_{gt} = (4.5)(Q_{da})(h_e - h_l)$$

Equation 9.14

where:

q_{gt} = grand total cooling load (Btuh)
Q_{da} = rate of conditioned airflow (cfm)
h_e = enthalpy of the entering (mixed) air (Btu/lb-dry air)
h_l = enthalpy of the leaving air (Btu/lb-dry air)
4.5 = constant to make units consistent

Example 9-5.

Using the air quantity, air conditions, and enthalpies of the air as it enters and leaves the cooling coil, what are the total and sensible cooling capacities for the system described in Example 9-4?

From the enthalpy scale on the upper left side of the psychrometric chart, Figure 9-8, and Equation 9.14:

$$q_{gt} = (4.5)(5,390 \ cfm) (30.4 \ Btu/lb \ dry \ air - 23.6 \ Btu/lb \ dry \ air) = 164,934 \ Btuh$$

From the entering and leaving dry bulb temperatures and Equation 6.1b:

$$q_S = (1.10)(5,390 \ cfm) (78.5°F - 56.5°F)$$
$$= 130,438 \ Btuh$$

In the above examples, the ratio of the sensible heat to the total heat was fairly high. The cooling load was mostly sensible heat. In some applications, this is not necessarily the case. In many applications, there is a comparatively larger latent cooling load. Examples of an area with large latent loads might include a gymnasium with a large latent cooling load from many occupants. Another might be a hospital operating room with an all outdoor air ventilation system.

In such instances, when the ESHF line is plotted on the psychrometric chart, the slope of the line is so steep that it either does not intersect the saturation curve at all or it intersects the curve at a ridiculously low temperature. The normal leaving air dry bulb temperature range for comfort air conditioning is about 48° to 55°F. Lower temperatures are difficult to achieve without special-application cooling coils. Higher leaving air temperatures will result in very little dehumidification.

In high latent cooling load applications, it is usually necessary to select a reasonable leaving air temperature and reheat the air before it enters the conditioned space. Reheating the air adds sensible heat to the leaving air, while the humidity ratio remains constant. It may also be necessary to provide a higher airflow rate in order to handle the latent cooling load.

Equation 9.15 may be used to estimate the amount of reheat required:

$$q_{rh} = \frac{ESHF[q_{rs} + q_{rl} + (BF)(q_{oas} + q_{oal})] - [q_{rs} + (BF)(q_{oag})]}{(1 - ESHF)}$$

Equation 9.15

where:

q_{rh} = heat required to reheat the dehumidified cooled air (Btuh)
ESHF = effective sensible heat factor
q_{rs} = room sensible cooling load (Btuh)

q_{rl} = room latent cooling load (Btuh)
q_{oas} = outdoor air sensible cooling load (Btuh)
q_{oal} = outdoor air latent cooling load (Btuh)
BF = coil bypass factor

Before using reheat in an hvac system, consult the applicable energy codes for the project location. Some energy codes prohibit or limit the use of reheat.

Example 9-6.
A gymnasium has a sensible cooling load of 110,000 Btuh and a latent load of 45,000 Btuh. The code requires 1,000 cfm of outdoor air for the system. The space design conditions are 75°F db, 50% rh; outdoor conditions are 94°F db, 75°F wb. The cooling coil has a bypass factor of 0.10.

What is the required total cooling capacity of the system? What is the required airflow rate? How much reheat is required (if any)? And what are the coil entering, leaving, and reheat discharge state points?

From Example 9-4:

q_{oas} = 20,900 Btuh
q_{oal} = 26,150 Btuh

Therefore:

q_{gt} = 110,000 Btuh + 45,000 Btuh + 20,900 Btuh + 26,150 Btuh = 202,050 Btuh

The bypass factor is 0.10, calculate the ESHF:

$$ESHF = \left(\frac{110,000 \text{ Btuh} + (0.10)(20,900 \text{ Btuh})}{\substack{110,000 \text{ Btuh} + (0.10)(20,900 \text{ Btuh}) \\ + 45,000 \text{ Btuh} + (0.10)(26,150 \text{ Btuh})}} \right) = 0.70$$

Plotting the ESHF = 0.70 on the psychrometric chart in Figure 9-9, it crosses the saturation curve at 46°F db. This is below the range of a normal cooling coil. Assume a more reasonable apparatus dewpoint of 50°F. Using a T_{adp} = 50°F db and plotting a line on the psychrometric chart results in ESHF = 0.75.

$$Q_{da} = \frac{110,000 \text{ Btuh} + (0.10)(20,900 \text{ Btuh})}{(1.10)(75°F - 50°F)(1 - 0.10)} = 4,528 \text{ cfm}$$

$$T_{mdb} = \frac{(3,530 \text{ cfm})(75°F) + (1,000 \text{ cfm})(94°F)}{4,530 \text{ cfm}} = 79°F \text{ db}$$

From the psychrometric chart in Figure 9-9:

T_{mwb} = 65.5°F
T_{ldb} = 50 + (0.10)(79°F - 50°F) = 52.9°F db
T_{lwb} = 52°F wb

From Equation 9.15:

$$q_{rh} = \frac{\substack{0.75[110,000 \text{ Btuh} + 45,000 \text{ Btuh} + \\ (0.10)(20,900 \text{ Btuh} + 26,150 \text{ Btuh})] - [110,000 \\ \text{Btuh} - (0.10)(20,900 \text{ Btuh})}}{(1 - 0.75)}$$

q_{rh} = 30,755 Btuh

Equation 6.1a may be rearranged to calculate a temperature difference. Calculating the temperature rise through the reheat coil:

$$T_2 - T_1 = \frac{30,755 \text{ Btuh}}{(1.08)(4,530)} = 6°F \text{ db}$$

The conditions leaving the reheat coil are 58.9°F db and 54.5°F wb. This gets the supply air conditions close to the original ESHF = 0.70 line.

Checking the total cooling capacity by getting the entering and leaving enthalpy values from the chart:

q_{gt} = (4.5)(4,530 cfm)(30.5 - 21.6) = 181,427 Btuh

Compared to the previous total cooling load, this value is about 11% lower. The calculated airflow for the above design conditions will handle the sensible cooling load; however, it will not handle the latent cooling load. Therefore, it is necessary to determine a new airflow rate that will satisfy both the sensible and latent cooling loads.

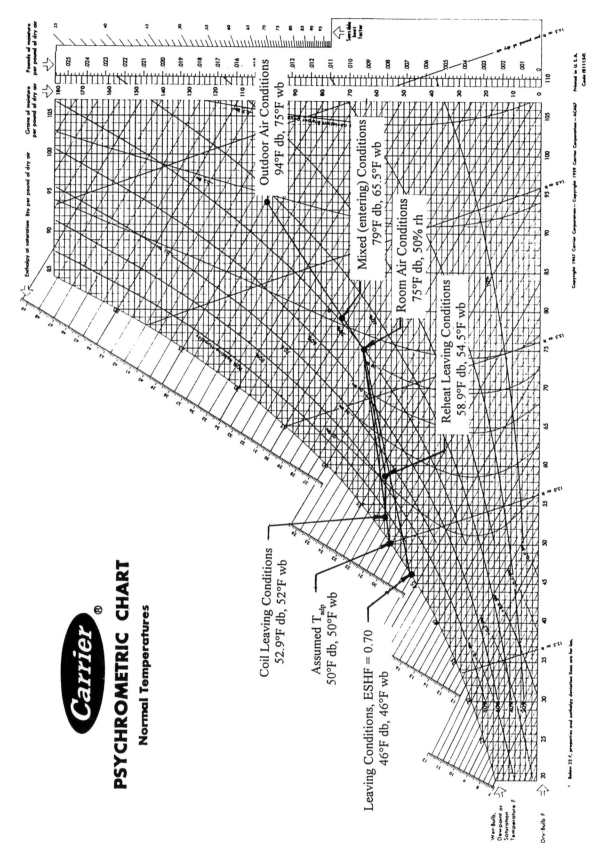

Carrier®

PSYCHROMETRIC CHART

Normal Temperatures

Outdoor Air Conditions
94°F db, 75°F wb

Mixed (entering) Conditions
79°F db, 65.5°F wb

Room Air Conditions
75°F db, 50% rh

Reheat Leaving Conditions
58.9°F db, 54.5°F wb

Coil Leaving Conditions
52.9°F db, 52°F wb

Assumed T$_{adp}$
50°F db, 50°F wb

Leaving Conditions, ESHF = 0.70
46°F db, 46°F wb

Figure 9-9. State points for Example 9-6.

Rearranging Equation 9.14 to solve for the airflow:

$$Q_{da} = \frac{202{,}050 \text{ Btuh}}{(4.5)(30.5 - 21.6)} = 5{,}045 \text{ cfm}$$

This airflow will handle the sensible and latent loads. Recalculating the reheat to get the required temperature rise:

$$q_{rh} = (1.10)(5{,}045 \text{ cfm})(6°F) = 33{,}297 \text{ Btuh}$$

The revised airflow rate will change the entering conditions slightly. It may be necessary to make several more calculations to get convergence.

Hvac psychrometric calculation procedure

This section describes the steps necessary to determine the final cooling loads for a system and to specify and select the hvac system for a space. It is assumed that the reader has already calculated the space cooling loads in accordance with Chapters 7 and 8.

To perform the psychrometric calculations for a system, it is necessary to have a psychrometric chart and several straight edges to draw lines. Blank psychrometric charts can be obtained from hvac equipment manufacturers and vendors. They are also available through ASHRAE.

1. Select the indoor design conditions for the project requirements and the outdoor air conditions for the project location from the climatic data in Chapter 3. Plot the state points on a psychrometric chart and draw a straight line between them.

2. Calculate the external, internal, and return cooling loads for the space in accordance with the procedures in Chapters 7 and 8.

3. Determine the outdoor ventilation air requirements. Use Equations 6.1b and 8.5 to calculate the outdoor air sensible and latent cooling loads.

4. Calculate the grand total cooling load (q_{gt}) using Equation 9.9.

5. Assume a value for the cooling coil bypass factor based on the expected application and the guidelines listed on page 119. Calculate the ESHF from Equation 9.11.

6. Locate the ESHF value on the SHF scale on the psychrometric chart. Draw a line from the ESHF on the SHF scale through the alignment circle near the center of the chart. (Other charts, such as the ASHRAE chart, use a slightly different method.) Draw a second line beginning at the room state point, parallel to the first line, extending to where it intersects the saturation curve on the chart. If the line does not intersect the saturation curve or it intersects the curve at a point beyond the range of normal hvac equipment, skip to Step 12.

7. The point where the second line intersects the saturation curve is the adp of the cooling coil. Read the dry bulb temperature from the psychrometric chart.

8. Use Equation 9.12 to calculate the conditioned air supply for the space.

9. Use Equation 9.5 to calculate the return air quantity and Equation 9.8 to calculate the cooling coil entering (mixed) air dry bulb temperature. Locate a point at the intersection of the entering dry bulb temperature and the line between the indoor and outdoor state points. Read the entering wet bulb temperature and the enthalpy from the chart. Draw a straight line from the entering air state point to the adp state point. This line represents the GSHF for the system.

10. Use Equation 9.13 to calculate the leaving dry bulb temperature. Locate a point at the intersection of the leaving dry bulb temperature and the GSHF line. Read the leaving wet bulb temperature and the enthalpy from the chart.

11. Use Equation 9.14 to calculate the total cooling load based on the conditioned air supply, entering enthalpy, and leaving

enthalpy. As a check, compare this value to the value calculated in Step 4 above. If they are in reasonable agreement, the psychrometric calculations for the total cooling load, conditioned air supply, and state points are correct.

12. If the ESHF line through the room state point does not intersect the saturation curve at a reasonable value, assume a reasonable value for the coil adp. Draw a straight line from the room state point through the assumed adp. Using the SHF scale on the chart, determine an equivalent ESHF.

13. Calculate the conditioned air supply, entering conditions, and leaving conditions, as described in Steps 8, 9, and 10 above, using the new ESHF line to locate the leaving conditions.

14. Using Equation 9.15, calculate the heat required for reheat (if any). Also calculate the resultant temperature rise in the air dry bulb temperature.

15. Draw a horizontal line from the cooling coil leaving conditions to the intersection of the reheat dry bulb temperature. This point locates the leaving conditions for the reheat coil. Read the reheat coil leaving wet bulb temperature from the chart.

Chapter 10

Overview of Hvac Systems

Chapter 1 of this book discussed some of the major requirements of hvac systems and some of the issues that should be considered when selecting an hvac system. Basic descriptions of system types are also included in Chapter 1. One major factor to consider when selecting an hvac system is the total heating, ventilating, and air conditioning capacity requirement of the system or systems. Chapters 5 through 9 provide the most widely accepted methods for determining various system capacity requirements. This chapter provides an overview of the hvac systems commonly used for heating, ventilating, and air conditioning buildings today. Methods for providing the heating and cooling energy also are discussed.

Basically, hvac systems can be grouped into three major categories: **air systems**, **hydronic-steam systems**, and **unitary** or **packaged systems**. Practically all systems can be classified as one of the major groups or a combination of these groups.

In central air systems, heating and cooling is accomplished by heating or cooling air that is circulated to the conditioned space. A conditioned "space" may be considered a group of thermally similar areas which, when grouped together, are called a zone. Each space with a different exposure, or with a different magnitude of heating or cooling load, may be considered a zone.

Each zone normally has a separate temperature control device (thermostat) and sometimes a humidity control device (humidistat). Zones have nearly uniform heating and/or cooling loads throughout the space. A zone is typically comprised of one or more rooms that are similar in their thermal load characteristics. A room implies an enclosed area that may or may not require a separate temperature control.

Central air systems

Chapter 1 gave a brief description of a basic air-handling system. Central air systems typically consist of a central air-handling unit (AHU) and an air distribution system of ductwork and air devices. The AHU typically has a fan, filters, a cooling coil, and sometimes a heating device. A typical AHU is shown in Figure 10-1.

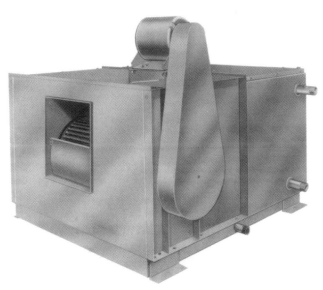

Figure 10-1. Typical air-handling unit. (Courtesy, York International Corporation)

Heating devices can include hot water heating coils, steam coils, electric coils, gas furnaces, and oil furnaces. Cooling coils are usually chilled water-type or direct expansion (DX) refrigerant coils. A typical water coil is shown in Figure 10-2.

Figure 10-3. Centrifugal fan. (Courtesy, The Trane Company)

Figure 10-2. Water heating or cooling coil. (Courtesy, York International Corporation)

AHUs in central air systems are usually "built-up" units; that is, they are comprised of various components that match the design requirements of the system. A built-up unit may have a chilled water cooling coil and an electric heating coil, or it may have a DX cooling coil and a heating water coil. In a built-up unit, it is usually possible to mix and match fan types, coil types, filter types, etc.

Fans in built-up AHUs are usually centrifugal fans. A centrifugal fan is shown in Figure 10-3.

Central air systems can basically be divided into three major types which describe how heating and cooling capacity control is achieved. The major types are **constant-volume, single-path**; **multi-path**; and **variable air volume (vav)** systems.

Constant-volume, single-path

The most basic type of central air hvac system is the constant-volume, single-path system. As the name implies, the airflow rate is constant, because there is only one path for the air to follow as it leaves the AHU and is distributed to the conditioned space. This type of system is best suited for a single zone, since the system is either in the heating or cooling mode at any given time. The system may be in the "dead band" mode, in which it is neither heating nor cooling, just circulating air.

It is not possible to simultaneously heat one zone and cool another with a single system. Typical applications for the constant-volume, single-path systems are single-family houses, large auditoriums with very few exterior surfaces, spaces with a single exterior exposure, and spaces with fairly uniform heating and/or cooling loads. A schematic diagram of a constant-volume, single-path system is shown in Figure 10-4.

Constant-volume, single-path systems normally have an AHU with heating and/or cooling coils and ductwork in a series flow path. Sometimes the heating coil may be replaced with a gas- or oil-fired furnace, as in the case of a residential system. The heating or cooling is cycled on or off, in response to a load in the conditioned space that is sensed by a thermostat.

A variation on the constant-volume, single-path system is the constant-volume reheat system. This system can provide either heating or cooling to multiple zones simultaneously with a constant air volume system. A schematic diagram of a constant-volume reheat system is shown in Figure 10-5.

In this type of system, air is circulated at a constant rate and leaves the AHU at a set temperature, usually about 55°F. Just before the air enters the conditioned zone, a duct-mounted reheat coil increases the temperature of the air entering the zone in accordance with the zone

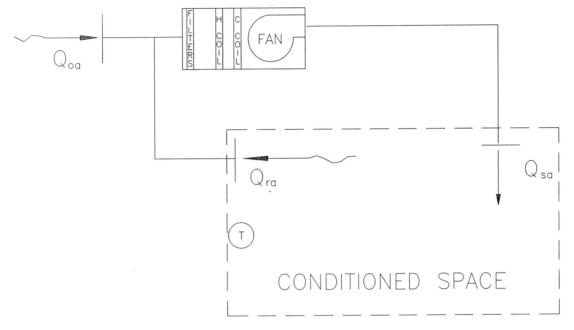

Figure 10-4. Constant-volume, single-path system schematic.

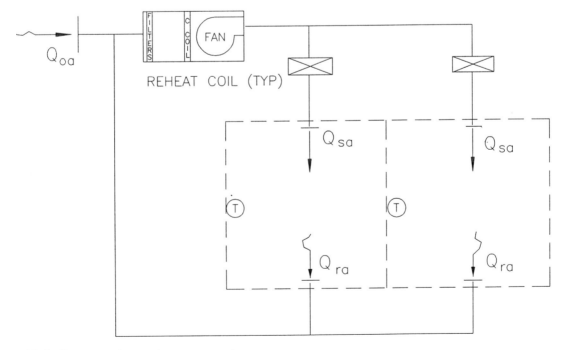

Figure 10-5. Constant-volume reheat system schematic.

heating or cooling load sensed by the zone thermostat. Reheat coils can be electric, steam, or heating water. A separate preheat coil may also be provided in the AHU to preheat the incoming outdoor air as necessary.

Constant-volume reheat systems are rarely used on new installations; they are very energy inefficient. Air is first cooled to a set supply air temperature and then heated to meet zone requirements. Some energy codes no longer allow constant-volume reheat systems.

Multi-path systems

As the name implies, multi-path systems allow the conditioned air to follow multiple parallel paths to each zone. The temperature of the air supplied to each zone is a function of the heating or cooling load in that zone. Typically, the AHU has both heating and cooling coils, in order to provide either heated or cooled air to a zone as required. The two most common multi-path air systems are the multi-zone system and the dual-duct system.

Multi-zone systems usually have an AHU with a cooling coil and multiple heating coils.

Each zone has its own supply air duct from the AHU, allowing the air to flow through multiple parallel supply air ducts to the zones. Each zone has its own reheat coil inside the AHU. The zone reheat coil is controlled by a zone thermostat.

Air is passed through the cooling coil in the AHU and is cooled to a predetermined temperature, usually about 55°F. The airflow is then divided inside the AHU into separate paths for each zone. Still inside the AHU, the air for each zone is then either passed through a reheat coil or bypassed around the reheat coil, according to the zone heating or cooling load. The proportion of air passed through, or around, the reheat coil is controlled by the zone thermostat.

Since the supply air is first cooled then reheated or bypassed to provide the desired mixed-air temperature, these systems are rather energy inefficient and wasteful. Very few multi-zone systems are installed in new buildings today.

Most multi-zone systems are found in older buildings.

The second multi-path system in use today is the dual-duct system. As the name implies, there are two ducts to each space or zone. One duct carries heated air while the other carries cooled air. There can be single or dual AHUs for the system. Either cooled air, heated air, or both are supplied to the space according to the thermal load in the space.

In order to provide air to the space at the required temperature, a dual-duct terminal is provided for each zone to mix the heated and cooled air in the required proportion. The dual-duct terminal mixes the two airstreams in accordance with the requirements of the zone thermostat. A typical dual-duct terminal is shown in Figure 10-6.

Notice that there are two air valves on the inlet of the terminal. One is for the cooled air, and the other is for heated air. The air valves throttle each airflow according to the requirements of the space thermostat.

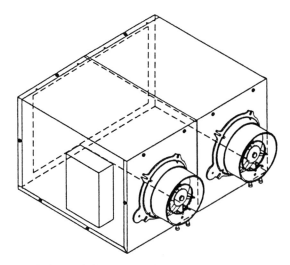

Figure 10-6. Dual-duct terminal unit. (Courtesy, The Trane Company)

Variable air volume

Other than the constant-volume, single-path system, the vav system is the most common type of air system installed in non-residential buildings today. As the name implies, the rate of airflow in the system is not constant, but varies in accordance with the total sensible cooling load for the building.

The supply air is usually cooled and supplied to the distribution ductwork at a constant temperature, although the air temperature may be reset in accordance with the cooling load. A schematic of a typical vav system is shown in Figure 10-7.

As may be observed from Figure 10-7, the vav system is also a single-path system. However, it is not constant air volume. The amount of cooled air admitted to the conditioned space is a function of the sensible cooling load in the space, which is sensed by a thermostat. As the temperature of the space decreases, an air valve in a vav terminal closes, reducing the airflow as the cooling load is satisfied.

Heat for the space is usually not provided by heating the supply air in the AHU. Frequently, heat is provided by baseboard radiators around the exterior perimeter of the conditioned space.

Heat may also be provided in the vav terminal with a reheat coil in the vav terminal itself. A typical vav terminal with a reheat coil is shown in Figure 10-8.

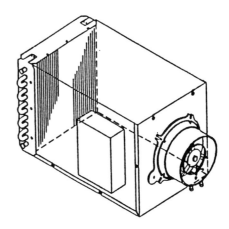

Figure 10-8. Vav terminal with hot water reheat. (Courtesy, The Trane Company)

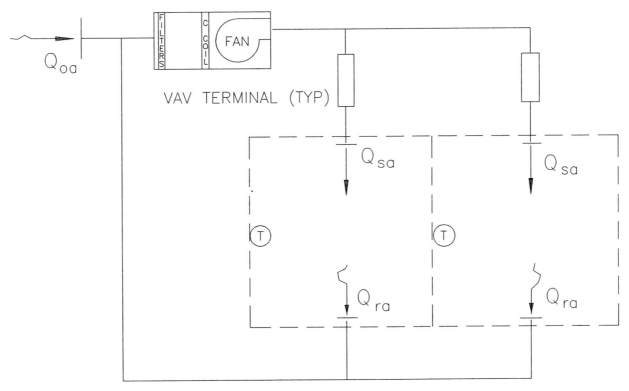

Figure 10-7. Vav system schematic.

Notice that an air valve is on the inlet of the vav terminal. The air valve throttles, or varies, the flow of conditioned air to the space in response to a space thermostat. For terminals with reheat, the reheat coil is usually located on the discharge of the terminal.

Since the quantity of air supplied to the space varies, it is good design practice to provide some method to control the airflow of the AHU fan. This is normally accomplished in one of three ways: fan discharge dampers, fan inlet vanes, or variable-frequency motor drives. Each method varies the flow of the AHU fan in response to changes in static air pressure in the main supply air duct.

The first, and least desirable, are discharge dampers. Discharge dampers are automatically actuated dampers that are placed on the outlet, or discharge, of the AHU. The dampers open and close in response to the pressure in the main supply air duct. As the vav terminals open and the pressure in the main duct drops, the discharge dampers open to allow more air from the fan into the ductwork.

As the vav terminals close and the duct pressure increases, the discharge dampers close. Discharge dampers are somewhat energy inefficient, since the fan is rotating at full speed while the dampers block off the airflow. Using discharge dampers is a little like driving an automobile at full throttle while controlling the speed with the brakes. Discharge dampers may also result in undesirable high air pressures in the AHU.

A variation on discharge dampers is the bypass system. Air is simply bypassed around the conditioned space and returned to the AHU. Air is normally "dumped" into a return air plenum or duct by a motorized relief damper. The fan in the AHU is still rotating at full speed, fully loaded. Bypass systems are typically found on smaller commercial air systems.

A second and more energy efficient method to control the AHU fan air volume is with inlet vanes on the fan. These vanes, located at the air inlet on the fan, open or close in response to the pressure in the main supply air duct. Figure 10-9 shows a centrifugal fan with inlet vanes.

Figure 10-9. Fan with inlet vanes. (Courtesy, York International Corporation)

Since the inlet vanes restrict the amount of air entering the fan, the fan rotates in an unloaded condition, even though it is rotating at full speed. This reduces the load on the electric motor driving the fan and reduces the power consumed by the motor. Inlet vanes are more costly than discharge dampers; however, they are also more effective.

The third and most effective way to control the airflow of a fan is a variable-frequency motor drive. These drives vary the frequency of the electric power to the fan motor in response to the pressure in the main supply air duct. The varying electrical frequency in turn causes the motor and fan to rotate at different speeds. When the air pressure in the duct drops, the fan speed increases to supply more air. When the air pressure increases, the fan speed decreases accordingly.

Although variable-frequency drives tend to be more expensive than the other methods to control air volume, they are also much more efficient. Frequently, the energy cost savings of a variable-frequency drive justifies the additional first cost.

Air distribution systems

In central air systems, conditioned air is usually distributed throughout a building to the conditioned spaces via a network of ductwork and air distribution devices. The ductwork and devices between the AHU and the conditioned space are called the **supply air system**; the devices and ductwork that return the air to the AHU are called the **return air system**.

Air distribution devices supply the air directly to, or return the air from, the conditioned space. Air devices include ceiling air diffusers, air registers, and grilles. In the design of hvac systems, the location of the air devices in the conditioned space is important. They should be located so as to provide a uniform air distribution throughout the space.

Supply air outlets and return inlets should not be located so close to each other that the airflow tends to "short circuit" from the outlet to the inlet. Air devices should also be located so as to not cause drafts within the space. However, the devices should be located where they will provide good air motion within the space and so that there are no areas of stagnant air within the space.

When selecting air devices for a particular application, care should be taken to ensure that the device will not cause an objectionable noise level in the space at the design airflow rate. Performance data for air devices normally includes NC (noise criteria) levels at various airflow rates. The higher the NC level, the higher the noise. Offices and similar spaces normally require a maximum NC level of around 30, while spaces such as restaurants or cocktail lounges, can have NC levels as high as 45.

Reference 3 lists the recommended NC levels for various types of spaces. The Air Distribution Council provides standard methods for rating the performance of air supply and return devices. NC data is usually provided with the air device manufacturer's literature.

In most central air systems, a network of ductwork is included to distribute the conditioned air from the AHU to the room devices. The ductwork system is actually two systems, a supply air system and a return air system. Sometimes much of the return air ductwork above a ceiling is not included in lieu of a ceiling return

air plenum. If a ceiling return air plenum is used, consult the applicable codes to determine which materials are permissible in return air plenums and for what applications return air plenums are allowed.

Supply air duct systems are usually classified according to their static air pressure and/or air velocity. Duct systems are classified as low pressure if the maximum static pressure is 2 in. wg or less. They are classified as low velocity if the air velocity in the ductwork is 2,000 fpm or less. Ductwork with pressures greater than 2 in. wg and/or air velocities greater than 2,000 fpm are considered high pressure, high velocity. The Sheet Metal and Air Conditioning Contractors' National Association (SMACNA) provides duct construction standards for ductwork handling air at various pressures and velocities.

There are a number of considerations involved when trying to make a decision as to whether a supply air duct system should be low velocity, low pressure or high velocity, high pressure. Some major considerations include the amount of air to be circulated, the space available for ductwork, noise level, and installed cost. Designers should pay particular attention to noise levels for ductwork routed through quiet or sensitive areas and in high-velocity systems. Reference 1 includes additional information on duct design.

Hydronic and steam systems

Hydronic systems provide heating and/or cooling to a space by circulating heated and/or chilled water. Heating water is usually generated by a boiler and circulated through a piping distribution system with a pump. Chilled water is usually produced by a chiller, a cooling tower, and circulating pumps. Chilled water and heating water systems are generally found in large buildings, building complexes, and buildings on a campus.

Hydronic systems are generally very energy efficient. However, the need for a central plant to generate chilled and/or heating water is usually a large part of the cost of an hvac system. Chillers, boilers, cooling towers, and related equipment are discussed later in this chapter.

In addition to the main central plant equipment, a hydronic system consists of one or more circulating pumps, piping, and some sort of heat exchange device. Heat exchangers transfer heat between the circulated water and the air in the space or the air supplied to the space by an AHU. Heat exchange devices, which include hydronic heating-cooling coils, fancoil units, unit heaters, and radiators are discussed later in this chapter.

Hydronic systems can have several piping arrangements. Usually, there is supply piping that distributes the circulated water from the central plant to the heat exchange devices, and return piping that returns the water to the primary heating or cooling device in the central plant. Such an arrangement is called a **two-pipe system**. When the heat exchange devices use both heating water and chilled water, there may be a **four-pipe system**: one set of pipes for the chilled water and one set for the heating water. Sometimes **three-pipe systems** are used; the return piping is a common return for both chilled and heating water.

Hydronic systems have a number of advantages. First, since water has a much higher specific heat than air, the cross-sectional area of a pipe required to carry water is much less than the cross-sectional area for a duct carrying air for any given thermal load. Therefore, a hydronic system requires much less space in occupied areas of the building than does an air system.

Second, hydronic systems also require less pumping horsepower than that required to circulate air for any given thermal load. This can result in lower operating costs.

Third, since the heat transfer device is frequently located in the conditioned space, hydronic systems are usually easier to zone if they are a four-pipe system.

Steam systems are similar to hydronic systems except that steam is distributed as the heating medium. Obviously, steam systems are not directly used for cooling. Circulating pumps are not required for steam heating systems, since the pressure of the steam itself is adequate for steam flow.

Steam condensate (steam that has condensed back to water) is usually returned via a gravity drainage system. Occasionally, the condensate is pumped if gravity drainage is not possible.

Steam systems are different from hydronic systems in that steam undergoes a phase change. As water is heated it is converted to steam, usually in a boiler. Steam is then converted back to water (condensate) in a heat exchange device, usually a steam heating coil or a steam radiator. Again it undergoes a phase change. The phase change releases the majority of the heat from steam.

For example, steam at 15 psig, 250°F will release about 945 Btu/lb of steam as it condenses from steam to condensate. The phase change occurs at a nearly constant temperature.

In order for the steam to condense in a heat exchange device, the steam itself must be kept inside the device until it condenses. After the steam becomes condensate, the condensate must be removed from the heat exchange device. This is usually accomplished with a steam trap. The trap will not allow steam to pass; however, it will allow condensate to pass. A steam trap is placed on the discharge piping of the heat exchange device to hold the steam in while allowing the condensate to flow out.

Heating and cooling coils

Typically, heating coils use either heating water or steam as the heating medium, although electric heating coils are also used. Cooling coils typically use chilled water or a refrigerant as the cooling medium. A typical water coil was shown in Figure 10-2. A typical DX type refrigerant coil is shown in Figure 10-10.

In all finned coils, some of the air is passed through the coil and is not affected by the coil (it is not heated or cooled). This air is called **bypass air**. The percentage of bypassed air can range from as high as 30% to as low as 2%. The amount of bypassed air depends on the air velocity, coil fin spacing, and the number of coil rows. Typical bypass factors for coils were given in Chapter 9.

Figure 10-10. DX refrigerant coil. (Courtesy, York International Corporation)

A coil is comprised of many fins (between which air flows), coil tubes, and headers. The fins increase the heat transfer area in contact with the passing air. Although much of the heat transfer from a coil is from the fins, the primary source of heat transfer is from round tubes, or pipes, that run between the coil headers at each end. The tubes are installed perpendicularly to the fins and airflow. If the fins run vertically, the tubes run horizontally. The number of tubes in the direction parallel to the airflow designates the number of rows of the coil.

The tubes are usually interconnected by return bends to form a serpentine arrangement. The fins are bonded to the tubes to provide good conduction heat transfer between the fluid in the tubes, the tubes themselves, and the fins. Water coils, steam coils, and refrigerant coils usually have aluminum fins and copper tubes. A header is usually provided at one or both ends of the tubes to connect and distribute the fluid to and from the tubes. Air vents and drains are usually provided on the headers of water coils.

DX refrigerant coils are a little more complicated than water coils. In a refrigerant coil, the refrigerant flows into the coil as a liquid and then evaporates to vapor as it absorbs heat from the

air. The refrigerant rate of flow must be properly metered and distributed to the tubes for uniform cooling throughout the coil. The flow of refrigerant into the cooling coil (or evaporator as it is sometimes called) is controlled by a thermostatic expansion valve at the inlet to the coil. DX refrigerant systems are discussed later in this chapter.

Fancoil units and unit ventilators

A fancoil unit is, as the name implies, a small unit with a fan and a chilled and/or heating water coil enclosed in a common cabinet. Fancoil units are usually small enough to be located in the space that they serve. The exterior cabinet frequently has a decorative finish. A typical fancoil unit is shown in Figure 10-11.

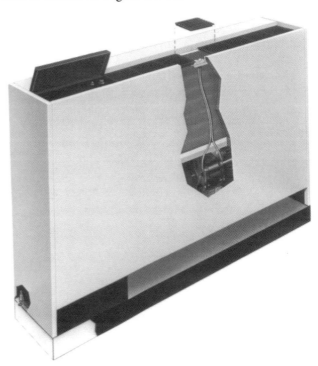

Figure 10-11. Fancoil unit. (Courtesy, The Trane Company)

Fancoil units typically have a small fan, fan motor, filters, a single water coil, and controls mounted inside a decorative cabinet. Fancoil units can also have electric heating coils for electrically heated units. A single water coil is usually used for both cooling and heating, even when the units are part of a four-pipe system. A control valve or valves mounted on the inlet to the water coil controls that set of pipes flow through the coil.

Fancoil units are usually located along exterior walls and can have limited outdoor air capabilities. The amount of outdoor air that a fancoil unit can handle is relatively small, since the cooling coils are designed as "dry coils." Dry coils cannot handle very large latent cooling loads, as would be the case with higher quantities of outside air.

Air is normally discharged vertically through a grille in the top of the unit and returned near the bottom front of the cabinet. An outdoor air connection (if any) is usually provided in the back of the unit that is against the wall. A drain pan is usually provided inside the cabinet under the coil to collect condensate from latent cooling. Fancoil units with bottom discharge and return are available for installations above ceilings.

Unit ventilators are similar in construction to fancoil units, except that they are designed to handle much larger outdoor air quantities and higher latent cooling loads. It is possible for a unit ventilator to handle up to 100% outside air. Unit ventilators frequently include modulating return air and outside air dampers to modulate the amount of outdoor air. Some unit ventilators have DX cooling coils as an alternative to a chilled water coil.

Unit heaters

As the name implies, unit heaters are small units that provide heating only. The heating medium for unit heaters is usually heating water or steam. Unit heaters can also have electric heating coils or be gas fired with gas furnaces. A typical unit heater is shown in Figure 10-12.

A unit heater consists of a heating element (a coil or furnace) and a small fan enclosed in a single cabinet. They can be horizontal or vertical discharge. Unit heaters are usually installed to provide heating in unfinished spaces such as warehouses and garages.

Figure 10-12. Unit heater. (Courtesy, The Trane Company)

Radiators

Radiators are frequently used for heating in both finished and unfinished areas of buildings. They are usually located against exterior walls and especially under windows. Radiators can use steam, heating water, or electricity as the heating medium.

Older buildings still use vertical sectional cast iron radiators. Although cast iron radiators are still in use today, they have been replaced by finned tube radiators in new buildings. The finned tube radiator has no moving parts. Steam and heating water radiators consist of a pipe through which the heating medium flows and metal fins bonded to the pipe. A finned tube element is shown in Figure 10-13.

The fins increase the heat transfer area of the heating element. Air flows through the fins by natural convection heat transfer. As the air is heated, it becomes less dense and rises. Air that rises is then replaced by cold air from below, which, in turn, is heated.

In finished areas, the finned tube elements are usually installed in decorative cabinets for aesthetic and safety reasons. Finned tube radiator cabinets are shown in Figure 10-14.

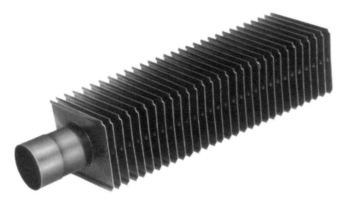

Figure 10-13. Finned tube element. (Courtesy, The Trane Company)

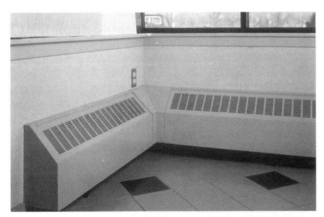

Figure 10-14. Finned tube radiator cabinets. (Courtesy, The Trane Company)

Unitary packaged heating and cooling units

Unitary or packaged hvac units are, as their name implies, a complete hvac unit in a single package or cabinet. Packaged units usually have a complete DX cooling system, fans, and sometimes heating. The cooling system includes a refrigerant compressor, evaporator coil, condenser coil, and condenser fan. Heating is usually a gas furnace or electric coil, although heating water and steam coils are also available. Packaged units are also available with a heat pump cycle for heating.

Packaged equipment is quite popular since it usually has a lower first cost. Since everything is factory built and installed in a single cabinet, equipment costs are comparatively low. Installation costs are usually low since it is usually only necessary to set the unit, connect the utilities, and connect the distribution ductwork (if any).

Packaged units, however, cannot provide very precise control of temperature and humidity. These units are primarily for basic comfort heating and cooling and are not suitable for process cooling or a situation with high latent cooling loads. Since these are packaged units, there are a limited number of options with the components. It is usually not possible to "build-up" a custom unit for a special application.

Window and through-the-wall units

Most people are familiar with window air conditioners and through-the-wall air conditioners. They are small-capacity units used primarily in residential applications, although they are commonly used in commercial applications for spot cooling and heating. Motel rooms frequently use through-the-wall units. Through-the-wall units are also referred to as packaged terminal air conditioners.

The units are usually completely self contained with all components in a single cabinet. Each unit is factory assembled with a DX cooling system and usually an electric coil for heat. They may also have reverse-cycle heat pump capabilities for heating. The units are complete, including all temperature controls.

Window and through-the-wall units usually have a single utility connection that is electric power. The evaporator fan has a low static pressure capability, therefore it should not be connected to ductwork. Window and through-the-wall units are available with cooling capacities from 0.5 to 7.5 tons.

Window and through-the-wall units must be mounted on an outside wall to allow the condenser to reject heat outdoors. The units usually have limited outdoor air ventilation capabilities.

Rooftop hvac units

As the name implies, rooftop hvac units are packaged units mounted on the roof of a building and ducted down through the roof to the spaces below.

Packaged rooftop units are also available with a horizontal duct discharge for grade-mounted units. Rooftop units are usually installed on a roof curb, which supports the unit and provides a weatherproof seal around the duct openings through the roof. Rooftop units are complete factory assembled units with a DX refrigeration system, heating system, fans, and controls in a single weatherproof cabinet. A typical rooftop unit is shown in Figure 10-15.

The most common forms of heating are gas furnaces and resistance electric heating coils. Units are also available with heating water coils and steam coils. Small- and medium-sized units are available as heat pumps.

Rooftop units are usually used in commercial applications such as office buildings, retail spaces, and other spaces requiring purely comfort cooling. They are quite popular because of their low first cost and the fact that they don't use any space inside the building. Usually, the only required connections to the unit are the supply-return ductwork, electrical power, and gas (if applicable).

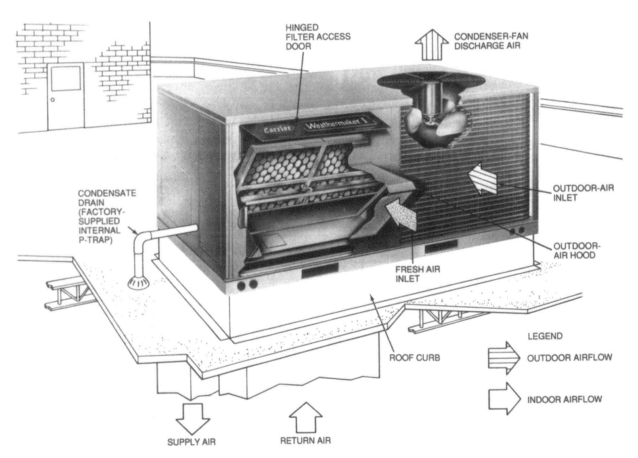

Figure 10-15. Rooftop hvac unit. (Courtesy, Carrier Corporation)

One of the major disadvantages to rooftop units is that they can cause considerable noise and vibration in the spaces near the unit, particularly under the unit.

Rooftop units are available with cooling capacities ranging from as low as 2.5 tons to as high as 125 tons in a single unit. Smaller units tend to be constant air volume; larger units can be variable air volume. Small- and medium-sized units can use an air bypass arrangement to achieve variable airflow to the spaces.

Heating methods and equipment

This section discusses the various methods or sources of heat and heating equipment for hvac systems. Heat can be generated within the conditioned space itself, in an AHU serving the space, or in a central plant that may provide heat for the entire building. Central plants may also serve many buildings, such as a campus.

Heat generated within the space is usually resistance electric heat and occasionally natural gas or liquid petroleum gas. Electric baseboard heaters and electric coils in air terminal units or packaged terminal hvac units are the most frequently used in space electric heating methods. Gas-fired unit heaters and similar equipment are the most common methods of providing gas heat from within the space. Of course, all gas-fired equipment of this type must be properly vented.

Heat may also be provided to the conditioned space by an AHU serving the space. Again, resistance electric coils, natural gas, and propane gas are common heat sources. Heat pumps and fuel oil may also be used as a heating energy source. In order to use gas or fuel oil, a furnace is usually used.

Heat can be generated by a central plant in the form of heating water or steam. For hydronic or steam heating, a boiler is used to heat the circulated water or to generate steam. Boilers can be electric, gas fired, or fuel oil fired. Some boilers can use gas or fuel oil (dual fuel).

Furnaces

Furnaces are frequently used for heating when the heating source is a fossil fuel (chemical energy) such as natural gas, liquid petroleum (lp) gas, or fuel oil. Furnaces typically include an air circulation fan, air filters, controls, a burner, and a heat exchanger in a single cabinet. Oil-fired furnaces frequently include a combustion air fan and motor (pressure atomizing burners). A typical furnace with an outdoor condensing unit (air conditioner) is shown in Figure 10-16.

Furnace controls usually include an ignition device (pilot light or igniter), gas valve, fan switch, and temperature limit switch. The fan switch turns the fan on when the heat exchanger reaches a set temperature, and turns off when the heat exchanger drops below the setpoint temperature. The temperature limit switch shuts the fuel off if the heat exchanger temperatures become excessive.

A heat exchanger is necessary for gas- and oil-fired furnaces to separate the combustion flue gases from the air circulated to the conditioned space. On older furnaces, air for the combustion process is drawn from the space in which the furnace is located. The flue gases are then discharged vertically through a flue pipe or chimney. On newer furnaces, a separate vent pipe is used to bring combustion air in directly from outdoors.

Furnaces are also used in commercial hvac systems, usually in packaged rooftop equipment and as duct furnaces for AHUs. Since it is difficult to provide zoning for large systems with furnaces, space heating for large buildings is usually accomplished with either hydronic or electric units located in the conditioned space. Furnaces in large rooftop units are usually for morning warmup after night setback.

Furnaces are usually rated according to their fuel consumption or input and their heating output capacity. The heating output in Btuh divided by the fuel input heating value in Btuh gives an approximation of the furnace's combustion efficiency. Older furnaces have a combustion efficiency around 70%; newer furnaces can have efficiencies as high as 95%.

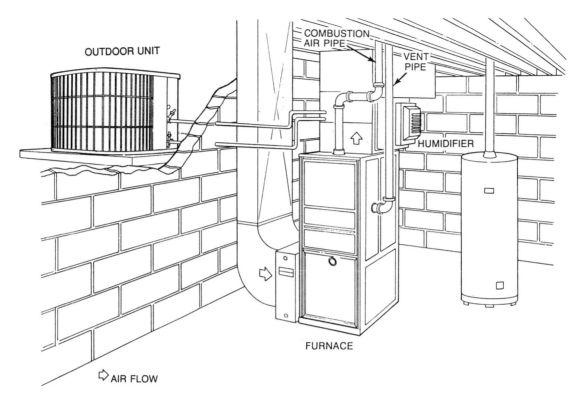

Figure 10-16. Residential furnace and air conditioner. (Courtesy, Carrier Corporation)

Electric heating

The most common forms of resistance electric heating are electric baseboard heaters, electric coils in AHUs, and electric coils in air terminal units. Typical applications for electric heating coils include outdoor air preheat and electric reheat in air terminal units.

Electric coils are ideal for preheat applications since they are not subject to freezing at low air temperatures. They are also well suited for air terminals, since they can readily be located in the conditioned space without the need for combustion venting or hydronic piping. Another form of electric heating is the heat pump, discussed later in this chapter.

Electric heating coils produce heat as the result of the resistance to the flow of electrical energy. For larger coils, over about 5 kW, it is necessary to energize the coil in stages for temperature control. It is usually not advisable to have a large coil come on all at once.

Safety controls for electric coils include primary and secondary thermal overload protection, overcurrent protection, disconnecting protection, and an airflow interlock. The airflow interlock ensures that air is flowing when the coil is energized to prevent overheating.

All electric heating equipment should be approved by Underwriters Laboratory and the National Electrical Code.

Heating boilers

Central heating plants use boilers as a source of heat to produce heating water or generate steam. The boilers can be gas-fired, fuel oil-fired, and occasionally electric. A typical gas-fired packaged steel boiler is shown in Figure 10-17.

Gas- and oil-fired boilers are similar to furnaces in that they have a burner and a heat exchanger. Burners can be either natural draft or power burners with a combustion air fan. Heat exchangers can be classified as fire tube or water tube.

Figure 10-17. Packaged steel boiler. (Courtesy, Kewanee Manufacturing Company)

Fire tube boilers have the hot combustion gases passing through the tubes surrounded by water. Water tube boilers, as the name implies, are the opposite of fire tube boilers. Water tube boilers have the water inside the tubes and the hot combustion gases flow around the tubes containing the water.

Steam boilers and systems are classified according to the system steam pressure. Low-pressure steam systems operate at steam pressures under 15 psig. Medium- and high-pressure steam systems operate at pressures greater than 15 psig. Steam boilers are typically available with steam capacities up to 50,000 lb of steam per hour.

Hydronic boilers and systems are classified according to their temperature and pressure. Low-temperature, low-pressure hydronic boilers operate up to 250°F and 160 psig pressure. Medium- and high-temperature, high-pressure hydronic systems operate at pressures greater than 160 psig and temperatures greater than 250°F.

Hydronic boilers are available with heating capacities up to 50,000,000 Btuh. Hydronic heating systems usually have a maximum operating temperature of 250°F or less and are usually designed for a water temperature drop of 20° to 40°F through a heating coil.

Both steam and heating water boilers are rated at their maximum working pressure and temperature as determined by the American Society of Mechanical Engineers (ASME) Boiler Code Section.

Boilers are usually constructed of cast iron or steel. Cast iron boilers are constructed of individual sections that can be bolted together to form the boiler. Steel boilers are usually packaged boilers that come from the factory completely assembled in a single welded steel jacket. Cast iron boilers range in capacity from 35,000 to 10,000,000 Btuh gross heat output. Packaged steel boilers start around 50,000 Btuh and go up to 50,000,000 Btuh gross heat output.

Operating and safety controls for boilers include temperature and pressure controls, temperature limit controls, high and low fuel pressure, flame failure fuel shut-off, pressure and temperature relief valve, and excess air control. Boilers must also have a "fuel train," which consists of regulators, gauges, valves, test cocks, pressure switches, etc., in the fuel piping to the boiler. Fuel trains are frequently furnished with the boiler.

Cooling and refrigeration systems

Most cooling and refrigeration systems operate as DX refrigeration systems. Some chillers operate on the absorption refrigeration principle. Absorption chillers are discussed in later in this chapter.

DX systems are classified as direct systems by ASHRAE Standard 15, *Safety Code for Mechanical Refrigeration.* A direct system is one in which the evaporator or condenser of the refrigerating system is in direct contact with the air to be cooled.

Chillers are classified as indirect systems in which a secondary coolant (such as water) is cooled by the refrigerating system, which is then circulated to the air to be cooled.

When selecting hvac equipment that has a refrigerating system and uses a refrigerant, the designer should be aware that some formerly popular refrigerants may no longer be available. The Montreal Protocol, which was adopted in 1987, has banned the production of some refrigerants (such as R-12 and R-502).

Although refrigeration equipment may continue to operate with these refrigerants, the refrigerants may not be available when the equipment needs servicing. When designing a cooling or heat pump system, consult the latest issue of ASHRAE Standard 15 and the latest applicable codes.

DX systems

The most common system for producing a refrigerating effect is the direct expansion (DX) system. DX systems operate on a thermodynamic cycle that transfers heat from a low-temperature region to a higher-temperature region. Since heat

only flows naturally from a high-temperature region to a low-temperature region, it is necessary to apply energy, in the form of work, to a refrigeration system. Work is provided by a compressor that compresses the refrigerant vapor. A simple DX refrigeration system is shown in Figure 10-18.

In the DX system, a refrigerant is circulated by the compressor. The refrigerant will undergo a phase change (from a gas to a liquid and back to a gas) as it goes through the cycle. The refrigerant enters the compressor as a gas and is compressed. Increasing the pressure of the gas causes its temperature to increase, and it becomes a superheated vapor. The superheated refrigerant vapor then passes through a condenser, where it rejects much of its heat.

The heat is usually rejected directly to the atmosphere or to a circulated cooling fluid, such as water. When the refrigerant comes out of the condenser, it is a saturated vapor at high pressure. (A high saturated substance is one in which vapor or liquid phases can be present at a given temperature and pressure.) The saturated vapor refrigerant then passes through an expansion valve.

The expansion valve reduces the pressure of the refrigerant, which also reduces its temperature. The refrigerant then becomes almost 100% saturated liquid at a much lower temperature. The saturated liquid refrigerant then passes through the evaporator.

In the evaporator, the liquid refrigerant evaporates and absorbs heat from the circulated air that is passed over the evaporator. The evaporator in an air conditioning system is the cooling coil. When the refrigerant comes out of the evaporator, it is again a vapor and the cycle repeats.

Figure 10-16 showed a residential heating and cooling system. This type of system is referred to as a split system. The evaporator coil and expansion valve are located in the furnace or AHU, while the compressor and condenser coil are in the outdoor unit. Refrigerant piping connects the two units and carries the refrigerant between them. The outdoor unit is usually called a condensing unit. A residential condensing unit is shown in Figure 10-19.

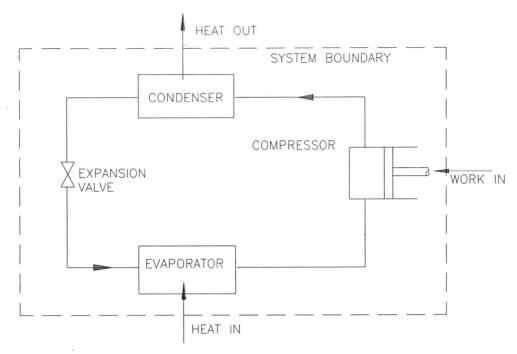

Figure 10-18. Simple DX refrigeration system schematic.

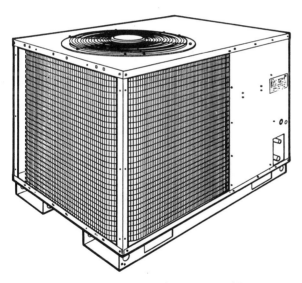

Figure 10-19. Residential condensing unit. (Courtesy, Carrier Corporation)

Residential and small commercial units have cooling capacities ranging from 2 to 6 tons. Large commercial condensing units can have cooling capacities over 150 tons. A large commercial condensing unit is shown in Figure 10-20.

When the outdoor unit contains the compressors, the condenser coils, and fans, it is referred to as a condensing unit. If the outdoor unit just has the condenser coils and fans with the compressors located elsewhere, the unit is referred to as a condenser.

A variation on the DX cooling system is the heat pump. A heat pump is used for heating a space instead of cooling. It is almost identical to the cooling system except that the direction of heat exchange is reversed. In winter, heat is transferred from outdoors to the indoors by the refrigeration process.

For a heat pump, the evaporator coil is outdoors and the condenser coil is indoors. Most heat pump systems function as a cooling system in summer and a heating system in winter. This is accomplished by a switching valve in the refrigerant piping that changes the function of the evaporator and condenser coils by changing the flow of the refrigerant.

Chilled water systems

Frequently, large buildings and multi-building campuses are cooled from a central plant. Central plants usually include one or more chillers to produce cooling, which is distributed to indi-

Figure 10-20. Large commercial condensing unit. (Courtesy, Carrier Corporation)

vidual air-handling and fancoil units by chilled water.

Chilled water systems typically supply chilled water between 42° and 50°F. If the water were any colder, it would require an anti-freezing solution and would be called a brine. Brine systems are generally only used for process cooling where lower temperatures are required.

Typically, the water in hvac systems is cooled by 8°, 10°, or 12°F as it passes through the chiller. As the chilled water passes through the chilled water coils in AHUs, the water temperature rises by an equivalent amount before it is returned to the chiller. The chilled water is circulated throughout the system by one or more chilled water pumps.

Most chillers operate on the same principle as the DX systems. The main difference is that the evaporator removes heat from the circulated water instead of air. Electrically driven chillers have a DX refrigeration system and are usually reciprocating, rotary screw, or centrifugal compressor chillers.

Reciprocating chillers use pistons inside of cylinders to compress the refrigerant. They typically have cooling capacities from 20 to as high as 200 tons. Rotary screw chillers use a rotating helical screw to compress the refrigerant. They have cooling capacities from 50 to 750 tons. Centrifugal chillers use a rotating wheel, or impeller, and centrifugal force to compress the refrigerant. Centrifugal chillers typically have cooling capacities ranging from 100 to 2,000 tons in a single machine. A centrifugal chiller is shown in Figure 10-21.

Chilled water may also be produced by absorption chillers, which do not operate on the DX cycle. Absorption chillers use heat instead of work from an electrical motor to produce a refrigerating effect. Most absorption chillers use a lithium bromide and water solution as the circulated refrigerant.

The cooling cycle is based on the absorption of heat from the chilled water when the refrigerant (water) evaporates. The refrigerant solution in the chiller is maintained at a near perfect vacuum.

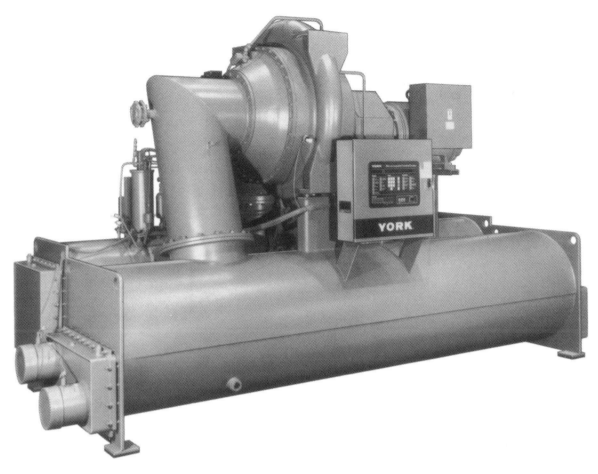

Figure 10-21. Centrifugal chiller. (Courtesy, York International Corporation)

The absorption cycle utilizes the affinity that lithium bromide has for water while under a vacuum.

The refrigerant (water) is cooled by latent cooling as it evaporates and is absorbed by a strong concentration of lithium bromide solution. As the lithium bromide solution becomes diluted by the water, it is transferred to another section, where it is heated. The heat causes the water to be driven off. The lithium bromide is then returned to the absorber to again absorb the water refrigerant. The water vapor that was driven out of the lithium bromide is condensed by cooling water and returned to the evaporator to repeat the cycle. Water is chilled as it is circulated through the tubes of the evaporator.

Smaller chillers, like reciprocating chillers, have condensers that are air cooled. Larger chillers, like centrifugal and absorption chillers, have condensers that are water cooled. Water-cooled

chillers reject heat to the atmosphere by a cooling tower, which is usually located outdoors. A typical cooling tower is shown in Figure 10-22.

Figure 10-22. Cooling tower. (Courtesy, Evapco, Inc.)

The amount of heat rejected by a cooling tower is increased if the circulated water comes in contact with the outdoor air. Most cooling towers spray water into the outside air as it is circulated through the cooling tower by fans. Spraying the water increases the evaporation of the water, which increases the heat rejection rate. As the water evaporates into the air, latent heat is transferred from the water to the air.

References

1. ASHRAE *Handbook—Fundamentals*, American Society of Heating, Refrigerating and Air-Conditioning Engineers, Inc., Atlanta, Georgia, 1989.

2. ASHRAE *Handbook—HVAC Applications*, American Society of Heating, Refrigerating and Air-Conditioning Engineers, Inc., Atlanta, Georgia, 1991.

3. ASHRAE *Handbook—HVAC Systems and Equipment*, American Society of Heating, Refrigerating and Air-Conditioning Engineers, Inc., Atlanta, Georgia, 1992.

4. ASHRAE GRP-158, *Cooling and Heating Load Calculation Manual*, American Society of Heating, Refrigerating and Air-Conditioning Engineers, Inc., Atlanta, Georgia, 1979.

5. Strock, C. and R.L. Koral, *Handbook of Air Conditioning, Heating and Ventilating*, Industrial Press, New York, 1965.

6. *Trane Air Conditioning Manual*, The Trane Company, La Crosse, Wisconsin, 1974.

7. *Carrier System Design Manual, Part 1, Load Estimating*, McGraw-Hill, Inc., New York, 1972.

8. Sontag, R.E. and G.J. Van Wylen, *Introduction to Thermodynamics, Classical and Statistical*, John Wiley and Sons, Inc., New York, 1971.

9. Albertson, M.L., J.R. Barton, and D.B. Simons, *Fluid Mechanics for Engineers*, Prentice-Hall, Inc., Englewood Cliffs, New Jersey, 1960.

10. Holman, J.P., *Heat Transfer*, McGraw-Hill, Inc., New York, 1972.

11. Chapman, A.J., *Heat Transfer*, Macmillan Publishing Co., Inc., New York, 1974.

12. Griffin, C.W. Jr., *Energy Conservation in Buildings: Techniques for Economical Design*, Construction Specification Institute, Washington, DC, 1974.

13. McQuiston, F.C., *Heating, Ventilating and Air Conditioning: Analysis and Design*, John Wiley and Sons, Inc., New York, 1977.

14. BOCA *National Mechanical Code*, Building Officials & Code Administrators International, Country Club Hills, Illinois, 1993.

15. Mull, T.E., *Design Procedure for a Hybrid Passive Energy Residence* (Master's thesis), University of Missouri, Rolla, Missouri, 1982.

Abbreviations, Symbols, and Conversions

air changes	AC	feet per minute	fpm
air-handling unit	AHU	feet per second	fps
apparatus dewpoint	adp	foot or feet	ft
approximate	approx	foot-pound	ft-lb
area	A	gallons	gal
bypass factor	BF	gallons per hour	gph
brake horsepower	Bhp	grand total sensible heat factor	GSHF
British thermal unit	Btu	grand total heat	GTH
conductance	C	heat transfer	q
conductivity	k	heat transfer coefficient	U
cubic feet	cu ft	horsepower	hp
cubic inch	cu in.	humidity, relative	rh
cubic feet per minute	cfm	humidity ratio	W
degree	degree or °	kilowatt	kW
diameter	dia, D	kilowatt hour	kWh
dry bulb temperature	db	leaving air temperature	LAT
effective room latent heat	ERLH	length	L
effective room sensible heat	ERSH	mass	m
effective sensible heat factor	ESHF	leaving water temperature	LWT
entering	ENT	maximum	max
entering air temperature	EAT	mercury	Hg
enthalpy	h	minimum	min
expansion	exp	noise criteria	NC
Fahrenheit	F	outside air	OA

percent	%		
pounds	lb		
pounds per square foot	lb/sq ft		
pounds per square inch	psi		
pressure	P, p		
Rankine	R		
relative humidity	rh		
return air	RA		
room latent heat	RLH		
room sensible heat	RSH		
room sensible heat factor	RSHF		
saturation	sat		
sensible heat	SH		
sensible heat ratio	SHR		
shading coefficient	SC		
specific heat	c		
square	sq		
static pressure	SP		
supply air	SA		
temperature	T		
thermal conductivity	k		
thermal resistance	R		
time	t		
U factor	U		
variable air volume	vav		
velocity	v		
vertical	vert		
volume	vol, V		
watt	W		
wet bulb temperature	wb		

Letter symbols

Symbol	Description	Typical units
A	area	sq ft
c	specific heat	Btu/lb-°F
C	thermal conductance	Btuh/sq ft-°F
D or dia	diameter	ft
h	enthalpy	Btu/lb
k	thermal conductivity	Btuh/sq ft-°F
p or P	pressure	psi
q	time rate of heat transfer	Btuh
Q	volumetric flow rate	cfm
r	radius	ft
R	thermal resistance	Btuh/sq ft-°F
T	temperature	°F
U	overall heat-transfer coefficient	Btuh/sq ft-°F
W	humidity ratio of moist air	lb (water)/ lb (dry air)
Δ	difference between values	
σ	Stefan-Boltzmann constant	Btuh/sq ft-°R^4
ϕ	relative humidity	%

T_{adp}	apparatus dewpoint temperature
T_{edb}	entering dry bulb temperature
T_{ewb}	entering wet bulb temperature
T_{ldb}	leaving dry bulb temperature
T_{lwb}	leaving wet bulb temperature
T_m	mixture of outdoor and return air

CONVERSION FACTORS

Multiply	By	To obtain
Atmospheres	29.92	Inches of mercury (at 32°F)
Atmospheres	33.97	Feet of water (at 62°F)
Atmospheres	14.697	Pounds per square inch
Boiler horsepower	33,475	Btu per hour
Btu	778	Foot-pounds
Btu	0.000293	Kilowatt hour
Btu per 24 hours	0.347×10^{-5}	Tons of refrigeration
Btu per hour	0.000393	Horsepower
Btu per hour	0.000293	Kilowatts
Btu per inch per square foot per hour per °F	0.0833	Btu per foot per sq ft per hour per °F
Cubic feet	1,728	Cubic inches
Cubic feet	7.48052	Gallons
Cubic feet of water	62.37	Pounds (at 60°F)
Feet of water	0.8826	Inches of mercury (at 32°F)
Feet of water	0.4335	Pounds per square inch
Feet per minute	0.01667	Feet per second
Foot-pounds	0.001286	Btu
Gallons (U.S.)	0.1337	Cubic feet
Gallons of water	8.3453	Pounds of water (at 60°F)
Horsepower	33,000	Foot-pounds per minute
Horsepower	2,547	Btu per hour
Horsepower	0.7457	Kilowatts
Inches of mercury (at 62°F)	0.03342	Atmospheres
Inches of mercury (at 62°F)	13.57	Inches of water (at 62°F)
Inches of mercury (at 62°F)	1.133	Feet of water (at 62°F)
Inches of mercury (at 62°F)	70.73	Pounds per square foot
Inches of mercury (at 62°F)	0.4912	Pounds per square inch
Inches of water (at 62°F)	0.07355	Inches of mercury
Inches of water (at 62°F)	0.03613	Pounds per square inch
Inches of water (at 62°F)	5.204	Pounds per square foot
Kilowatts	1.341	Horsepower
Kilowatt hours	3415	Btu
Pounds	7,000	Grains
Pounds of water (at 60°F)	0.01602	Cubic feet
Pounds of water (at 60°F)	27.68	Cubic inches
Pounds of water (at 60°F)	0.1198	Gallons
Pounds per square inch	0.06804	Atmospheres
Pounds per square inch	2.307	Feet of water (at 62°F)
Pounds per square inch	2.036	Inches of mercury (at 62°F)
Temperature °C + 273	1	Absolute temperature (K)
Temperature °C + 17.78	1.8	Temperature (°F)
Temperature °F + 460	1	Absolute temperature (R)
Temperature °F - 32	5/9	Temperature (°C)
Tons of refrigeration	12,000	Btu per hour
Watts	3.415	Btu per hour
Watt-hours	3.415	Btu

Thermal Properties of Selected Materials

Thermal Properties of Building Materials

Description	Thickness	Thermal Properties			
		Conductivity	Density	Specific Heat	Resistance
	ft	Btuh/sq ft-°F	lb/cu ft	Btu/lb-°F	sq ft-°F/Btuh
Acoustic Tile					
3/8 inch	0.0313	0.0330	18.0	0.32	0.95
1/2 inch	0.0417	0.0330	18.0	0.32	1.26
3/4 inch	0.0625	0.0330	18.0	0.32	1.89
Aluminum or Steel Siding	0.005	26	480	0.1	0
Asbestos - Cement					
1/8 inch board	0.0104	0.3450	120.0	0.2	0.03
1/4 inch board	0.0208	0.3450	120.0	0.2	0.06
Shingle					0.21
1/4 inch Lapped Siding					0.21
Asbestos - Vinyl Tile				0.3	0.05
Asphalt					
Roofing Roll			70.0	0.35	0.15
Shingle and Siding			70.0	0.35	0.44
Tile				0.3	0.05
Brick					
4 inch Common	0.3333	0.4167	120.0	0.2	0.80
8 inch Common	0.6667	0.4167	120.0	0.2	1.60
12 inch Common	1.0000	0.4167	120.0	0.2	2.40
3 inch Face	0.2500	0.7576	130.0	0.22	0.33
4 inch Face	0.3333	0.7576	130.0	0.22	0.44
Built-up Roofing, 3/8 inch	0.03	0.09	70	0.35	0.33
Building Paper					
Permeable Felt					0.06
2-Layers Seal					0.12
Plastic Film Seal					0.01
Carpet					
With Fibrous Pad				0.34	2.08
With Rubber Pad				0.34	1.23
Cement					
1 inch Mortar	0.0833	0.4167	116.0	0.2	0.20
1-3/4 inch Mortar	0.1458	0.4167	116.0	0.2	0.35
1 inch Plaster with Sand Aggregate	0.0833	0.4167	116.0	0.2	0.20

Thermal Properties of Building Materials

Description	Thickness	Thermal Properties			
		Conductivity	Density	Specific Heat	Resistance
	ft	Btuh/sq.ft-°F	lb/cu.ft	Btu/lb-°F	sq.ft-°F/Btuh
Clay Tile, Hollow					
3 inch 1 cell	0.2500	0.3125	70.0	0.2	0.80
4 inch 1 cell	0.3333	0.2999	70.0	0.2	1.11
6 inch 2 cells	0.5000	0.3300	70.0	0.2	1.52
8 inch 2 cells	0.6667	0.3600	70.0	0.2	1.85
10 inch 2 cells	0.8333	0.3749	70.0	0.2	2.22
12 inch 3 cells	1.0000	0.4000	70.0	0.2	2.50
Clay Tile, Paver					
3/8 inch	0.0313	1.0416	120.0	0.2	0.03
Concrete, Heavy Weight Dried Aggregate, 140 lb					
1-1/4 inch	0.1042	0.7576	140.0	0.2	0.14
2 inch	0.1667	0.7576	140.0	0.2	0.22
4 inch	0.3333	0.7576	140.0	0.2	0.44
6 inch	0.5000	0.7576	140.0	0.2	0.66
8 inch	0.6667	0.7576	140.0	0.2	0.88
10 inch	0.8333	0.7576	140.0	0.2	0.10
12 inch	1.0000	0.7576	140.0	0.2	1.32
Concrete, Heavy Weight Undried Aggregate, 140 lb					
3/4 inch	0.0625	1.0417	140.0	0.2	0.06
1-3/8 inch	0.1146	1.0417	140.0	0.2	0.11
3-1/4 inch	0.2708	1.0417	140.0	0.2	0.26
4 inch	0.3333	1.0417	140.0	0.2	0.32
6 inch	0.5000	1.0417	140.0	0.2	0.48
8 inch	0.6667	1.0417	140.0	0.2	0.64
Concrete Block, 12 inch Heavy Weight					
Hollow	1.000	0.7813	76.0	0.2	1.28
Concrete Filled	1.000	0.7575	140.0	0.2	1.32
Partially Filled Concrete	1.000	0.7773	98.0	0.2	1.29
Concrete Block 4 inch Medium Weight					
Hollow	0.3333	0.3003	76.0	0.2	1.11
Concrete Filled	0.3333	0.4456	115.0	0.2	0.75
Perlite Filled	0.3333	0.1512	78.0	0.2	2.20
Partially Filled Concrete	0.3333	0.3306	89.0	0.2	1.01
Concrete and Perlite	0.3333	0.2493	90.0	0.2	1.34

Thermal Properties of Building Materials

Description	Thickness	Thermal Properties			
		Conductivity	Density	Specific Heat	Resistance
	ft	Btuh/sq.ft-°F	lb/cu.ft	Btu/lb-°F	sq.ft-°F/Btuh
Concrete Block, 6 inch Medium Weight					
Hollow	0.5000	0.3571	65.0	0.2	1.40
Concrete Filled	0.5000	0.4443	119.0	0.2	1.13
Perlite Filled	0.5000	0.1166	67.0	0.2	4.29
Partially Filled Concrete	0.5000	0.3686	83.0	0.2	1.36
Concrete and Perlite	0.5000	0.2259	84.0	0.2	2.21
Concrete Block, 8 inch Medium Weight					
Hollow	0.6667	0.3876	53.0	0.2	1.72
Concrete Filled	0.6667	0.4957	123.0	0.2	1.34
Perlite Filled	0.6667	0.1141	56.0	0.2	5.84
Partially Filled Concrete	0.6667	0.4648	76.0	0.2	1.53
Concrete and Perlite	0.6667	0.2413	77.0	0.2	2.76
Concrete Block, 12 inch Medium Weight					
Hollow	1.000	0.4959	58.0	0.2	2.02
Concrete Filled	1.000	0.4814	121.0	0.2	2.08
Partially Filled Concrete	1.000	0.4919	79.0	0.2	2.03
Concrete, Light Weight, 80 lb.					
3/4 inch	0.0625	0.2083	80.0	0.2	0.30
1-1/4 inch	0.1042	0.2083	80.0	0.2	0.50
2 inch	0.1667	0.2083	80.0	0.2	0.80
4 inch	0.3333	0.2083	80.0	0.2	1.60
6 inch	0.5000	0.2083	80.0	0.2	2.40
8 inch	0.6667	0.2083	80.0	0.2	3.20
Concrete Light Weight, 30 lb.					
3/4 inch	0.0625	0.0751	30.0	0.2	0.83
1-1/4 inch	0.1042	0.0751	30.0	0.2	1.39
2 inch	0.1667	0.0751	30.0	0.2	2.22
4 inch	0.3333	0.0751	30.0	0.2	4.44
6 inch	0.5000	0.0751	30.0	0.2	6.66
8 inch	0.6667	0.0751	30.0	0.2	8.88
Concrete Block, 4 inch Heavy Weight					
Hollow	0.3333	0.4694	101.0	0.2	0.71
Concrete	0.3333	0.7575	140.0	0.2	0.44
Perlite Filled	0.3333	0.3001	103.0	0.2	1.11
Partially Filled Concrete	0.3333	0.5844	114.0	0.2	0.57
Concrete and Perlite	0.3333	0.4772	115.0	0.2	0.70

Thermal Properties of Building Materials

Description	Thickness	Thermal Properties			
		Conductivity	Density	Specific Heat	Resistance
	ft	Btu/sq.ft-°F	lb/cu.ft	Btu/lb-°F	sq.ft-°F/Btuh
Concrete Block, 6 inch Heavy Weight					
Hollow	0.5000	0.5555	85.0	0.2	0.90
Concrete Filled	0.5000	0.7575	140.0	0.2	0.66
Perlite Filled	0.5000	0.2222	88.0	0.2	2.25
Partially Filled Concrete	0.5000	0.6119	104.0	0.2	0.82
Concrete and Perlite	0.5000	0.4238	104.0	0.2	1.18
Concrete Block, 8 inch Heavy Weight					
Hollow	0.6667	0.6060	69.0	0.2	1.10
Concrete Filled	0.6667	0.7575	140.0	0.2	.88
Perlite Filled Concrete	0.6667	0.2272	70.0	0.2	2.93
Partially Filled Concrete	0.6667	0.6746	93.0	0.2	0.99
Concrete and Perlite	0.6667	0.4160	93.0	0.2	1.60
Concrete Block, 4 inch Light Weight					
Hollow	0.3333	0.2222	65.0	0.2	1.50
Concrete Filled	0.3333	0.3695	104.0	0.2	0.90
Perlite Filled Concrete	0.3333	0.1271	67.0	0.2	2.62
Partially Filled Concrete	0.3333	0.2808	78.0	0.2	1.19
Concrete and Perlite	0.3333	0.2079	79.0	0.2	1.60
Concrete Block, 6 inch Light Weight					
Hollow	0.5000	0.2777	55.0	0.2	1.80
Concrete Filled	0.5000	0.3819	110.0	0.2	1.31
Perlite Filled	0.5000	0.0985	57.0	0.2	5.08
Partially Filled Concrete	0.5000	0.3189	73.0	0.2	1.57
Concrete and Perlite	0.5000	0.1929	74.0	0.2	2.59
Concrete Block, 8 inch Light Weight					
Hollow	0.6667	0.3333	45.0	0.2	2.00
Concrete Filled	0.6667	0.4359	115.0	0.2	1.53
Perlite Filled	0.6667	0.0963	48.0	0.2	6.92
Partially Filled Concrete	0.6667	0.3846	68.0	0.2	1.73
Concrete and Perlite	0.6667	0.2095	69.0	0.2	3.18
Concrete Block, 12 inch Light Weight					
Hollow	1.0000	0.4405	49.0	0.2	2.27
Concrete Filled	1.0000	0.4194	113.0	0.2	2.38
Partially Filled	1.0000	0.4274	70.0	0.2	2.34

Thermal Properties of Building Materials

Description	Thickness	Thermal Properties			
		Conductivity	Density	Specific Heat	Resistance
	ft	Btuh/sq.ft-°F	lb/cu.ft	Btu/lb-°F	sq.ft-°F/Btuh
Gypsum or Plaster Board					
1/2 inch	0.0417	0.0926	50.0	0.2	0.45
5/8 inch	0.0521	0.0926	50.0	0.2	0.56
3/4 inch	0.0625	0.0926	50.0	0.2	0.67
Gypsum Plaster					
3/4 inch Light Weight Aggregate	0.0625	0.1330	45.0	0.2	0.47
1 inch Light Weight Aggregate	0.0833	0.1330	45.0	0.2	0.63
3/4 inch Sand Aggregate	0.0625	0.4736	105.0	0.2	0.13
1 inch Sand Aggregate	0.0833	0.4736	105.0	0.2	0.18
Hard Board, 3/4 inch					
Medium Density Siding	0.0625	0.0544	40.0	0.28	1.15
Medium Density Others	0.0625	0.0608	50.0	0.31	1.03
High Density Standard Tempered	0.0625	0.0687	55.0	0.33	0.92
High Density Service Tempered	0.0625	0.0833	63.0	0.33	0.75
Linoleum Tile				0.3	0.05
Particle Board					
Low Density, 3/4 inch	0.0625	0.0450	75.0	0.31	1.39
Medium Density, 3/4 inch	0.0625	0.7833	75.0	0.31	0.08
High Density, 3/4 inch	0.0625	0.9833	75.0	0.31	0.06
Underlayment, 5/8 inch	0.0521	0.1796	75.0	0.29	0.29
Plywood					
1/4 inch	0.0209	0.0667	34.0	0.29	0.31
3/8 inch	0.0313	0.0667	34.0	0.29	0.47
1/2 inch	0.0417	0.0667	34.0	0.29	0.63
5/8 inch	0.0521	0.0667	34.0	0.29	0.78
3/4 inch	0.0625	0.0667	34.0	0.29	0.94
1 inch	0.0833	0.0667	34.0	0.29	1.25
Roof Gravel or Slag					
1/2 inch	0.0417	0.8340	55.0	0.4	0.05
1 inch	0.0833	0.8340	55.0	0.4	0.10
Rubber Tile					0.05
Slate, 1/2 inch	0.0417	0.834	100	0.35	0.05
Stone, 1 inch	0.0833	1.0416	140	0.2	0.08
Stucco, 1 inch	0.0833	0.4167	166	0.02	0.2
Terrazzo, 1 inch	0.0833	1.0416	140	0.2	0.08

Thermal Properties of Building Materials

Description	Thickness	Thermal Properties			
		Conductivity	Density	Specific Heat	Resistance
	ft	Btuh/sq.ft-°F	Btu/lb-°F	Btu/lb-°F	sq.ft-°F/Btuh
Wood, Soft					
3/4 inch	0.0625	0.0667	32.0	0.33	0.94
1-1/2 inch	0.1250	0.0667	32.0	0.33	1.87
2-1/2 inch	0.2083	0.0667	32.0	0.33	3.12
3-1/2 inch	0.2917	0.0667	32.0	0.33	4.37
4 inch	0.3333	0.0667	32.0	0.33	5.00
Wood, Hard					
3/4 inch	0.0625	0.0916	45.0	0.30	0.68
1 inch	0.0833	0.0916	45.0	0.30	0.91
Wood, Shingle					
For Wall	0.0583	0.0667	32.0	0.3	0.87
For Roof					0.94

Thermal Properties of Insulating Materials

Description	Thickness	Thermal Properties			
		Conductivity	Density	Specific Heat	Resistance
	ft	Btu/sq.ft-°F	lb/cu.ft	Btu/lb-°F	sq.ft-°F/Btuh
Mineral Wool/Fiberglass					
Batt, 2-1/2 inch, R-7	0.1882	0.0250	6.0	0.2	7.53
Batt, 3-1/2 inch, R-11	0.2957	0.0250	6.0	0.2	11.83
Batt, 6 inch, R-19	0.5108	0.0250	6.0	0.2	20.43
Batt, 8-1/2 inch, R-24	0.6969	0.0250	6.0	0.2	27.88
Batt, 9-1/2 inch, R-30	0.8065	0.0250	6.0	0.2	32.26
Fill, 3-1/2 inch, R-11	0.2917	0.0270	6.0	0.2	10.80
Fill, 5-1/2 inch, R-19	0.4583	0.0270	6.3	0.2	16.97
Cellulose					
Fill, 3-1/2 inch, R-13	0.2917	0.0225	3.0	0.33	12.96
Fill, 5-1/2 inch, R-20	0.4583	0.0225	3.0	0.33	20.37
Preformed Mineral Board					
7/8 inch, R-3	0.0729	0.0240	15.0	0.17	3.04
1 inch, R-3.5	0.0833	0.0240	15.0	0.17	3.47
2 inch, R-6.9	0.1667	0.0240	15.0	0.17	6.95
3 inch, R-10.3	0.2500	0.0240	15.0	0.17	10.42
Polystyrene, Expanded					
1/2 inch	0.0417	0.0200	1.8	0.29	2.08
3/4 inch	0.0625	0.0200	1.8	0.29	3.12
1 inch	0.0833	0.0200	1.8	0.29	4.16
1-1/4 inch	0.1042	0.0200	1.8	0.29	5.21
2 inch	0.1667	0.0200	1.8	0.29	8.33
3 inch	0.2500	0.0200	1.8	0.29	12.50
4 inch	0.3333	0.0200	1.8	0.29	16.66
Polyurethane, Expanded					
1/2 inch	0.0417	0.0133	1.5	0.38	3.14
3/4 inch	0.0625	0.0133	1.5	0.38	4.67
1 inch	0.0833	0.0133	1.5	0.38	6.26
1-1/4 inch	0.1042	0.0133	1.5	0.38	7.83
2 inch	0.1667	0.0133	1.5	0.38	12.53
3 inch	0.2500	0.0133	1.5	0.38	18.80
4 inch	0.3333	0.0133	1.5	0.38	25.06
Urea Formaldehyde					
3-1/2 inch, R-19	0.2910	0.0200	0.7	0.3	14.55
5-1/2 inch, R-30	0.4580	0.0200	0.7	0.3	22.90

Thermal Properties of Insulating Materials

Description	Thickness	Thermal Properties			
		Conductivity	Density	Specific Heat	Resistance
	ft	Btuh/sq.ft-°F	lb/cu.ft	Btu/lb-°F	sq.ft-°F/Btuh
Insulation Board					
Sheathing, 1/2 inch	0.0417	0.0316	18.0	0.31	1.32
Sheathing, 3/4 inch	0.0625	0.0316	18.0	0.31	1.98
Shingle Dacker, 3/8 inch	0.0313	0.0331	18.0	0.31	0.95
Nail Base Sheathing, 1/2 inch	0.0417	0.0366	25.0	0.31	1.14
Roof Insulation, Preformed					
1/2 inch	0.0417	0.0300	16.0	0.2	1.39
1 inch	0.0833	0.0300	16.0	0.2	2.78
1-1/2 inch	0.1250	0.0300	16.0	0.2	4.17
2 inch	0.1667	0.0300	16.0	0.2	5.56
2-1/2 inch	0.2083	0.0300	16.0	0.2	6.94
3 inch	0.2500	0.0300	16.0	0.2	8.33

Thermal Properties of Air Films and Air Spaces

Description	Thickness	Thermal Properties			
		Conductivity	Density	Specific Heat	Resistance
	ft	Btuh/sq.ft-°F	lb/cu.ft	Btu/lb-°F	sq.ft-°F/Btuh
Air Layer, 3/4 inch or less					
Vertical Walls					0.90
Slope 45 degrees					0.84
Horizontal Roofs					0.82
Air Layer, 3/4 inch to 4 inch					
Vertical Walls					0.89
Slope 45 degrees					0.87
Horizontal Roofs					0.87
Air Layer, 4 inch or more					
Vertical Walls					0.92
Slope 45 degrees					0.89
Horizontal Roofs					0.92

Index

A

Angle of incidence 70

B

Boilers 142, 143
Bypass factor 119

C

Chilled water system 145
Codes 4, 41
   ASHRAE Standard 62 5, 42
   BOCA 42
Condensation 109
Control volume 10
Cooling load 57, 59, 87, 115
   calculation 82, 102
   equipment 93, 95, 97
   estimating 83, 102
   exterior surfaces 59
   factors 59
   fan-motor 100
   infiltration 98
   lighting 90, 92
   outdoor air 117
   people 87
   return-air plenum 101
   roofs 66
   system 100
   temperature differences 59, 63
   unconditioned spaces 82
   walls 60
   windows and glass 74

D

Dalton's Law 108
Darcy's Equation 18
Design temperatures 19
Direct expansion system 144
Dry bulb temperature 109, 112

E

Electronic filters 5
Energy 10
Energy conservation 2
Enthalpy 11, 110
Evaporation 108

F

Fluid mechanics 16
Friction losses 17
Furnaces 141
   electric 142

H

Heat gain 57
   calculation methods 58
   direct 101
   external 58
   internal 87
   shading 72, 74
   windows and glass 68, 71
Heat loss 48
Heat transfer 12
   coefficient 33, 59, 66
   conduction 12

convection 13
    radiation 14
    rate 38
    transient 15
Heating load 48, 51
Heating methods 141
Hvac systems 5
    air 6, 129, 130, 132, 134
    fancoil unit 137, 138
    hydronic and steam 6, 135, 136
    radiators 138
    rooftop units 140
    selection of 7
    unit heaters 138
    unitary 7, 139
    variable air volume 132
    window units 139

I

Ideal Gas Law 108
Indoor air quality 4, 5
Infiltration 41, 43, 44, 49, 50, 98

L

Latent heat 2, 9, 11, 49, 88
    effective 120
Law of Conservation 17

N

National Institute of Occupational Safety and Health 4

P

Pressure 16
Psychrometrics 107, 110
    applying 117
    chart 110
    hvac calculation 127

R

Radiant heat transfer 2
Refrigeration 144
Relative humidity 109

S

Sensible heat 2, 9, 11, 50, 87
    factor 120
Sling psychrometer 110
Solar intensity 32
Specific heat 9, 11
Stack effect 42, 43

T

Thermal mass 36
Thermodynamics 10
    First Law 10, 49

U

U value. See Heat transfer: coefficient

V

Ventilation 41, 49, 50
Viscosity 17

W

Wet bulb temperature 109, 112
    depression 110
Wind effect 43

Other Titles Offered by BNP